TEXTBOOK OF BRYOLOGY

TEXTBOOK OF BRYOLOGY

Dr. P.R. Yadav
Lecturer
Dept. of Zoology
D.A.V. College
Muzaffarnagar (U.P.)
(India)

DISCOVERY PUBLISHING HOUSE PVT. LTD.
NEW DELHI-110 002

First Published - 2010

Reprinted - 2015

ISBN: 978-81-8356-577-6

Textbook of Bryology

Published by:

DISCOVERY PUBLISHING HOUSE PVT. LTD.
4383/4B, Ansari Road, Darya Ganj
New Delhi-110 002 (India)
Phone: +91-11-23279245, 43596064-65
Fax: +91-11-23253475
E-mail: discoverypublishinghouse@gmail.com
sales@discoverypublishinggroup.com
web: www.discoverypublishinggroup.com

Printed at:
Infinity Imaging Systems
Delhi

Preface

Most plants that people tend to think of are vascular (they contain internal tubes for transporting food and water within them) such as most house plants, flowering plants, etc. Bryophytes have poorly developed tubes-they are non-vascular plants. And, instead of reproducing with seeds, bryophytes reproduce using spores.

Because of the way bryophytes transport food and water, and because they require water for reproduction, bryophytes are small plants, and are most often abundant in wet places. However, many are quite drought tolerant and occur on trunks of trees, in forest canopies, and on dry rock surfaces.

Moss stems may be branched and have green leaves that are usually one cell thick. Often there is a bundle of cells (costa) present.

Moss leaves vary greatly in appearance, (including shiny, dull; red, green, or yellow), when they are moist or dry, and often change in orientation.

Mosses, liverworts and hornworts are found throughout the world in a variety of habitats. They flourish particularly well in moist, humid forests like the fog forests of the Pacific northwest or the montane rain forests of the southern hemisphere. Their ecological roles are many.They provide seed beds for the larger plants of the community, they capture and recycle nutrients that are washed with rainwater from the canopy and they bind the soil to keep it from eroding. In the northern hemisphere peatlands, wetlands often dominated by the moss *Sphagnum*, are particularly important bryophyte communities. This moss has exceptional water-holding capacity, and when dried and compressed, forms a coal-like fuel. Throughout northern Europe, Asia and North America, peat has been harvested for centuries for both fuel consumption and horticultural uses and today peatlands are managed as a sustainable resource.

—Author

Contents

Preface

1. **Introduction** 1

2. **Bryophyte** 8
   Introduction
   Non-vascular Plant
   Mosses
   Takakia
   Sphagnum
   Polytrichum Commune
   The Life Cycle of Mosses
   Monoicous Organisms
   Moss Reproduction
   Reproduction and Dispersal
   Sexual vs. Vegetative
   Sexual Reproduction
   Vegetative Reproduction
   The Importance of Mosses
   The Sporophyte

3. **Bryophyta** 81
   Introduction
   Sporangium
   Peristome
   The Fascinating Bryophyte

4. Hornworts **88**
Introduction
Ceratophyllum
Alternation of Generations
Evolutionary History of Plants

5. Marchantiophyta **124**
Introduction
Spore
Endospore
Mosses and Liverworts

6. Earlier Land Plants **143**
Introduction
Silurian-Devonian Terrestrial Algae and Problematica
Rhynia
Lush Life

7. Evolution and Diversity **156**
Introduction
The First Arthropods
The Hexapods
Apterygote Insects
Blattodea
Orthoptera
The Hemipteroid Group
Hemiptera
The Endopterygote Group

8. Lycopodium **201**
Introduction
Lycopodiopsida
Order Lycopodiales
Fern Ally
Vascular Plant

Index *211*

1

Introduction

Bryology is the study of mosses, liverworts, and hornworts (otherwise known collectively as Bryophytes).

Bryophytes are green land plants; there are approximately 15,000 species (9000 mosses, 5-6000 liverworts).

Most plants that people tend to think of are vascular (they contain internal tubes for transporting food and water within them) such as most house plants, flowering plants, etc. Bryophytes have poorly developed tubes - they are non-vascular plants. And, instead of reproducing with seeds, bryophytes reproduce using spores

Because of the way bryophytes transport food and water, and because they require water for reproduction, bryophytes are small plants, and are most often abundant in wet places. However, many are quite drought tolerant and occur on trunks of trees, in forest canopies, and on dry rock surfaces.

Moss stems may be branched and have green leaves that are usually one cell thick. Often there is a bundle of cells (costa) present.

Moss leaves vary greatly in appearance, (including shiny, dull; red, green, or yellow), when they are moist or dry, and often change in orientation.

Most liverworts are leafy (like mosses) but a few are thalloid (composed of a flattened body - the thallus - that is not differentiated into stem and leaves).

Leafy liverworts have 2 or 3 rows of leaves on their stems. The leaves never have a costa and often they are rounded or have 2-4 points at their tips, and are folded. If present, the leaves on the underside of the stem (the underleaves) are often reduced and usually pressed flat to the stem. Because the underleaves are reduced or absent, the larger stem leaves have a flattened appeareance, an arrangement that is rare in mosses.

Hornworts are thalloid and have linear "horn-like" capsules that split into two valves when mature. The cells of the thallus usually have only one chloroplast per cell - a condition not found in mosses or liverworts.

Plant scientists recognize two kinds of land plants, namely, bryophytes, or nonvascular land plants and tracheophytes,or vascular land plants. Bryophytes are small, herbaceous plants that grow closely packed together in mats or cushions on rocks, soil, or as epiphytes on the trunks and leaves of forest trees. Bryophytes are distinguished from tracheophytes by two important characters. First, in all bryophytes the ecologically persistent, photosynthetic phase of the life cycle is the haploid, gametophyte generation rather than the diploid sporophyte; bryophyte sporophytes are very short-lived, are attached to and nutritionally dependent on their gametophytes and consist of only an unbranched stalk, or seta, and a single, terminal sporangium. Second, bryophytes never form xylem tissue, the special lignin - containing, water-conducting tissue that is found in the sporophytes of all vascular plants. At one time, bryophytes were placed in a single phylum, intermediate in position between algae and vascular plants. Modern studies of cell ultrastructure and molecular biology, however,confirm that bryophytes comprise three separate evolutionary lineages, which are today recognized as mosses (phylum Bryophyta), liverworts (phylum Marchantiophyta) and hornworts (phylum Anthocerotophyta). Following a detailed analysis of land plant relationships, the three groups of bryophytes represent a grade or structural level in plant evolution, identified by their "monosporangiate" life cycle. Within this the geologically oldest group, sharing a fossil record with the oldest vascular plants in the Devonian era.

Of the three phyla of bryophytes, greatest species diversity is found in the mosses, with up to 15,000 species recognized. A moss begins its life cycle when haploid spores, which are produced in the sporophyte capsule, land on a moist substrate and begin to germinate. From the one-celled spore, a highly branched system of filaments, called the protonema, develops. Cell specialization occurs within the protonema to form a horizontal system of reddish-brown, anchoring filaments, called caulonemal filaments and upright, green filaments, called chloronemal filaments. Each protonema, which superficially resembles a filamentous alga, can spread over several centimeters to form a fuzzy green film over its substrate. As the protonema grows, some cells of the caulonemal filaments specialize to form leafy buds that will ultimately form the adult gametophyte shoots. Numerous shoots typically develop from each protonema so that, in fact, a single spore can give rise to a whole clump of moss plants. Each leafy shoot continues to grow apically, producing leaves in spiral arrangement on an elongating stem. In many mosses the stem is differentiated into a central strand of thin-walled water-conducting cells, called hydroids, surrounded by a parenchymatous cortex and a thick-walled epidermis. The leaves taper from a broad base to a pointed apex and have lamina that are only one-cell layer thick. A hydroid-containing midvein often extends from the stem into the leaf. Near the base of the shoot, reddish-brown, multicellular rhizoids emerge from the stem to anchor the moss to its substrate. Water and mineral nutrients required for the moss to grow are absorbed, not by the rhizoids,but rather by the thin leaves of the plant as rain water washes through the moss cushion.

As is typical of bryophytes, mosses produce large, multicellular sex organs for reproduction. Many bryophytes are unisexual, or sexually dioicous. In mosses male sex organs, called antheridia, are produced in clusters at the tips of shoots or branches on the male plants and female sex organs, the archegonia, are produced in similar fashion on female plants. Numerous motile sperm are produced by mitosis inside the brightly colored, club-shaped antheridia while a single egg develops in the base of each vase-shaped archegonium. As the sperm mature, the antheridium swells and bursts open. Drops of rain water falling into the cluster of open antheridia splash the sperm to near-by females. Beating their two whiplash flagellae, the sperm are able to move short distances in the water film that covers the plants to the open necks of the archegonia. Slimey mucilage secretions in the archegonial neck help pull the sperm downward to the egg. The closely packed arrangement of the individual moss plants greatly facilitates fertilization. Rain forest bryophytes that hang in long festoons from the trees rely on torrential winds with the rain to transport their sperm from tree to tree, while the small pygmy mosses of exposed, ephemeral habitats depend on the drops of morning dew to move their sperm.Regardless of where they grow, all bryophytes require water for sperm dispersal and subsequent fertilization.

Embryonic growth of the sporophyte begins within the archegonium soon after fertilization. At its base, or foot, the growing embryo forms a nutrient transfer zone, or placenta, with the gametophyte. Both organic nutrients and water move from the gametophyte into the sporophyte as it continues to grow. In mosses the sporophyte stalk, or seta, tears the archegonial enclosure early in development, leaving only the foot and the very base of the seta embedded in the gametophyte. The upper part of the archegonium remains over the tip of the sporophyte as a cap-like calyptra. Sporophyte growth ends with the formation of a sporangium or capsule at the tip of the seta. Within the capsule, water-resistant spores are formed by meiosis. As the mature capsule swells, the calyptra falls away. This allows the capsule to dry and break open at its tip. Special membranous structures, called peristome teeth, that are folded down into the spore mass, now bend outward, flinging the spores into the drying winds. Moss spores can travel great distances on the winds, even moving between continents on the jet streams. Their walls are highly protective, allowing some spores to remain viable for up to 40 years. Of course, if the spore lands in a suitable, moist habitat, germination will begin the cycle all over again.

Liverworts and hornworts are like mosses in the fundamental features of their life cycle, but differ greatly in organization of their mature gametophytes and sporophytes. Liverwort gametophytes can be either leafy shoots or flattened thalli. In the leafy forms, the leaves are arranged on the stem in one ventral and two lateral rows or ranks, rather than in spirals like the mosses. The leaves are one cell layer thick throughout, never have a midvein and are usually divided into two or more parts called lobes. The ventral leaves, which actually lie against the substrate, are usually much smaller than the lateral leaves and are hidden by the stem. Anchoring rhizoids, which arise near the ventral leaves, are colorless and unicellular. The flattened ribbon-like to leaf-like thallus of the thallose liverworts can be either simple or structurally differentiated into a system of dorsal air chambers and ventral storage tissues. In the latter type, the dorsal epidermis of the thallus is punctuated with scattered pores that open into the air chambers. Liverworts synthesize a vast array of volatile oils, which they store in unique organelles called oil bodies. These compounds impart an often spicy aroma to the plants and seem to discourage animals from feeding on them. Many of these compounds have potential as antimicrobial or anticancer pharmaceuticals.

Liverwort sporophytes develop completely enclosed within gametophyte tissues until their capsules are ready to open. The seta, which is initially very short,consists of small, thin-walled, hyaline cells. Just prior to capsule opening, the seta cells lengthen, thereby increasing the length of the seta upto 20 times its original dimensions. This rapid elongation pushes the darkly pigmented capsule and upper part of the whitish seta out of the gametophytic tissues. With drying, the capsule opens by splitting into four segments, or valves. The spores are dispersed into the winds by the twisting motions of numerous intermixed sterile cells, called elaters. In contrast to mosses, which disperse their spores over several days, liverworts disperse the entire spore mass of a single capsule in just a few minutes.

Hornworts resemble some liverworts in having simple, unspecialized thalloid gametophytes, but they differ in many other characters. For example, colonies of the symbiotic cyanobacterium *Nostoc* fill small cavities that are scattered throughout the ventral part of the hornwort thallus. When the thallus is viewed from above, these colonies appear as scattered blue-green dots. The cyanobacterium converts nitrogen gas from the air into ammonium, which the hornwort requires in its metabolism and the hornwort secretes carbohydrate- containing mucilage which supports the growth of the cyanobacterium. Hornworts also differ from all other land plants in having only one large, algal-like chloroplast in each thallus cell. Hornworts get their name from their long, horn-shaped sporophytes. As in other bryophytes, the sporophyte is anchored in the gametophyte by a foot through which nutrient transfer from gametophyte to sporophyte occurs. The rest of the sporophyte, however, is actually an elongate sporangium in which meiosis and spore development take place. At the base of the sporangium, just above the foot, is a mitotically active meristem,which adds new cells to the spore-producing zone throughout the life span of the sporophyte. In fact, the sporangium can be releasing spores at its apex, at the same time that new spores are being produced by meiosis at its base. Spore release in hornworts takes place gradually over a long period of time, and the spores are mostly dispersed by water movements rather than by wind.

Mosses, liverworts and hornworts are found throughout the world in a variety of habitats. They flourish particularly well in moist, humid forests like the fog forests of the Pacific northwest or the montane rain forests of the southern hemisphere. Their ecological roles are many. They provide seed beds for the larger plants of the community, they capture and recycle nutrients that are washed with rainwater from the canopy and they bind the soil to keep it from eroding. In the northern hemisphere peatlands, wetlands often dominated by the moss *Sphagnum*, are particularly important bryophyte communities. This moss has exceptional water-holding capacity, and when dried and compressed, forms a coal-like fuel. Throughout northern Europe, Asia and North America, peat has been harvested for centuries for both fuel consumption and horticultural uses and today peatlands are managed as a sustainable resource.

2

Bryophyte

INTRODUCTION

Bryophytes are all embryophytes ('land plants') that are non-vascular: they have tissues and enclosed reproductive systems, but they lack vascular tissue that circulates liquids. They neither have flowers nor produce seeds, reproducing via spores. The term bryophyte comes from Greek - bruon, "tree-moss, oyster-green" - bruo, "to be full to bursting, to abound" - fyton "plant".

The bryophytes do not form a monophyletic group but consist of three groups, the Marchantiophyta (liverworts), Anthocerotophyta (hornworts), and Bryophyta (mosses).

Modern studies of the land plants generally show one of two patterns. In one of these patterns, the liverworts were the first to diverge, followed by the hornworts, while the mosses are the closest living relatives of the polysporangiates (which include the vascular plants). In the other pattern, the hornworts were the first to diverge, followed by the vascular plants, while the mosses are the closest living relatives of the liverworts. Originally the three groups were brought together as the three classes of division Bryophyta. However, since the three groups of bryophytes form a paraphyletic group, they now are placed in three separate divisions.

These plants are generally gametophyte-oriented; that is, the normal plant is the haploid gametophyte, with the only diploid structure being the sporangium in season. As a result, bryophyte sexuality is very different from that of other plants. There are two basic categories of sexuality in bryophytes:

- Dioicous bryophytes produce only antheridia (male organs) or archegonia (female organs) on a single plant body.
- Monoicous bryophytes produce both antheridia and archegonia on the same plant body.

Some bryophyte species may be either monoicous or dioicous depending on environmental conditions. Other species grow exclusively with one type of sexuality.

Notice that these terms are not the same as monoecious and dioecious, which refer to whether or not a sporophyte plant bears one or both kinds of gametophyte. Those terms apply only to seed plants.

The embryophytes are the most familiar group of plants. They include trees, flowers, ferns, mosses, and various other green land plants. All are complex multicellular eukaryotes with specialized reproductive organs. With very few exceptions, embryophytes obtain their energy through photosynthesis (that is, by absorbing light); and they synthesize their food from carbon dioxide. Embryophyta may be distinguished from chlorophyll-using multicellular algae by having sterile tissue within the reproductive organs. Furthermore, embryophytes are primarily adapted for life on land, although some are secondarily aquatic. Accordingly, they are often called land plants or terrestrial plants.

Embryophytes developed from complex green algae (Chlorophyta) during the Paleozoic era. The Charales or stoneworts appear to be the best living illustration of that developmental step. These alga-like plants undergo an alternation between haploid and diploid generations (respectively called gametophytes and sporophytes). In the first embryophytes, however, the sporophytes became very different in structure and function, remaining small and dependent on the parent for their entire brief life. Such plants are informally called 'bryophytes'. They include three surviving groups:

- Bryophyta (mosses)
- Anthocerotophyta (hornworts)
- Marchantiophyta (liverworts)

All of the above 'bryophytes' are relatively small and are usually confined to moist environments, relying on water to disperse their spores. Other plants, better adapted to terrestrial conditions, appeared during the Silurian period. During the Devonian period, they diversified and spread to many different land environments, becoming the vascular plants or tracheophytes. Tracheophyta have vascular tissues or tracheids, which transport water throughout the body, and an outer layer or cuticle that resists drying out. In most vascular plants, the sporophyte is the dominant individual, and develops true leaves, stems, and roots, while the gametophyte remains very small.

Many vascular plants, however, still disperse using spores. They include two extant groups:

- Lycopodiophyta (clubmosses)
- Pteridophyta (ferns, whisk ferns, and horsetails)

Other groups, which first appeared towards the end of the Paleozoic era, reproduce using desiccation-resistant capsules called seeds. These groups are accordingly called spermatophytes or seed plants. In these forms, the gametophyte is completely reduced, taking the form of single-celled pollen and ova, while the sporophyte begins its life enclosed within the seed. Some seed plants may even survive in extremely arid conditions, unlike their more water-bound precursors. The seed plants include the following extant groups:

- Cycadophyta (cycads)
- Ginkgophyta (ginkgo)

- Pinophyta (conifers)
- Gnetophyta (gnetae)
- Magnoliophyta (flowering plants)

The first four groups are referred to as gymnosperms, since the embryonic sporophyte is not enclosed until after pollination. In contrast, among the flowering plants or angiosperms, the pollen has to grow a tube to penetrate the seed coat. Angiosperms were the last major group of plants to appear, developing from gymnosperms during the Jurassic period, and then spreading rapidly during the Cretaceous. They are the predominant group of plants in most terrestrial biomes today.

Note that the higher-level classification of plants varies considerably. Some authors have restricted the kingdom Plantae to include only embryophytes, others have given them various names and ranks. The groups listed here are often considered divisions or phyla, but have also been treated as classes, and they are occasionally compressed into as few as two divisions. Some classifications, indeed, consider the term Embryophyta at the superphylum (superdivision) level, and include Land Plants and some Charophyceae in a subkingdom named Streptophyta.

On a microscopic level, embryophyte cells remain very similar to those of green algae. They are eukaryotic, with a cell wall composed of cellulose and plastids surrounded by two membranes. These usually take the form of chloroplasts, which conduct photosynthesis and store food in the form of starch, and characteristically are pigmented with chlorophylls a and b, generally giving them a bright green color. Embryophytes also generally have an enlarged central vacuole or tonoplast, which maintains cell turgor and keeps the plant rigid. They lack flagella and centrioles except in certain gametes.

NON-VASCULAR PLANT

Non-vascular plants is a general term for those plants (including the green algae) without a vascular system (xylem and phloem). Although non-vascular plants lack these particular tissues, a number of non-vascular plants possess tissues specialized for internal transport of water.

Non-vascular plants have no roots, stems, or leaves, since each of these structures is defined by containing vascular tissue. The lobes (rounded parts) of the liverwort may look like leaves, but they are not true leaves because they have no xylem or phloem. Likewise, mosses and algae have no such tissues.

All plants have a life cycle with an alternation of generations between a diploid sporophyte and a haploid gametophyte, but nonvascular plants include the only plants that have a dominant gametophyte generation. In these plants, the sporophytes grow attached and are dependent on gametophytes for taking in water and other materials. Non-vascular plants grow from spores.

The term non-vascular plant is no longer used in scientific nomenclature. Non-vascular plants include two distantly related groups:

- *Bryophytes*— the Bryophyta (mosses), the Marchantiophyta (liverworts), and the Anthocerotophyta (hornworts). In these groups, the primary plants are the haploid gametophytes, with the only diploid portion being the attached sporophyte, consisting of a stalk and sporangium. Because these plants lack the water-conducting tissues, they fail to achieve the structural complexity and size of most vascular plants.
- *Algae*— especially the green algae. Recent studies have demonstrated that the algae actually consist of several unrelated groups. It turns out that common features of living in water and photosynthesis were misleading as indicators of close relationship. Only the green algae are still considered relatives of the plants..

Both of these groups are occasionally termed the "lower plants"; the term "lower" refers to these plants' status as the earliest plants to evolve. However, the term "lower" plants is not precise, since it frequently is used to include some vascular plants, the ferns and fern allies.

In the past, the term non-vascular plant included all the algae, but also the fungi as well. Today, it is recognized that these groups are not closely related to plants, and have a very different biology.

MOSSES

Mosses are small, soft plants that are typically 1-10 cm (0.4-4 in) tall, though some species are much larger. They commonly grow close together in clumps or mats in damp or shady locations. They do not have flowers or seeds, and their simple leaves cover the thin wiry stems. At certain times mosses produce spore capsules which may appear as beak-like capsules borne aloft on thin stalks.

There are approximately 12,000 species of moss classified in the Bryophyta The division Bryophyta formerly included not only mosses, but also liverworts and hornworts. These other two groups of bryophytes now are often placed in their own divisions.

Botanically, mosses are bryophytes, or non-vascular plants. They can be distinguished from the apparently similar liverworts (Marchantiophyta or Hepaticae) by their multi-cellular rhizoids. Other differences are not universal for all mosses and all liverworts, but the presence of clearly differentiated "stem" and "leaves", the lack of deeply lobed or segmented leaves, and the absence of leaves arranged in three ranks, all point to the plant being a moss.

In addition to lacking a vascular system, mosses have a gametophyte-dominant life cycle, i.e. the plant's cells are haploid for most of its life cycle. Sporophytes (i.e. the diploid body) are short-lived and dependent on the gametophyte. This is in contrast to the pattern exhibited by most "higher" plants and by most animals. In seed plants, for example, the haploid generation is represented by the pollen and the ovule, whilst the diploid generation is the familiar flowering plant.

Most kinds of plants excluding algae and bryophytes have two sets of chromosomes in their vegetative cells and are said to be diploid, i.e. each chromosome has a partner that contains the same, or similar, genetic information. Mosses (and other bryophytes) have only a single set of chromosomes (haploid, i.e. each chromosome exists in a unique copy within the cell). There are periods in the moss lifecycle when they do have a full, double set of paired chromosomes but this is only during the sporophyte stage.

The life of a moss starts from a haploid spore, which germinates to produce a protonema, which is either a mass of filaments or thalloid (flat and thallus-like). This is a transitory stage in the life of a moss. From the protonema grows the gametophore ("gamete-bearer") that is differentiated into stems and leaves. From the tips of stems or branches develop the sex organs of the mosses. The female organs are known as archegonia (sing. archegonium) and are protected by a group of modified leaves known as the perichaetum (plural, perichaeta). The archegonia have necks called venters which the male sperm swim down. The male organs are known as antheridia (singular antheridium) and are enclosed by modified leaves called the perigonium (plural, perigonia).

Mosses can be either dioicous (compare dioecious in seed plants) or monoicous (compare monoecious). In dioicous mosses, both male and female sex organs are borne on different gametophyte plants. In monoicous (also called autoicous) mosses, they are borne on the same plant. In the presence of water, sperm from the antheridia swim to the archegonia and fertilisation occurs, leading to the production of a diploid sporophyte. The sperm of mosses is biflagellate, i.e. they have two flagellae that aid in propulsion. Since the sperm must swim to the archegonium, fertilisation cannot occur without water. After fertilisation, the immature sporophyte pushes its way out of the archegonial venter. It takes about a quarter to half a year for the sporophyte to mature. The sporophyte body comprises a long stalk, called a seta, and a capsule capped by a cap called the operculum. The capsule and operculum are in turn sheathed by a haploid calyptra which is the remains of the archegonial venter. The calyptra usually falls off when the capsule is mature. Within the capsule, spore-producing cells undergo meiosis to form haploid spores, upon which the cycle can start again. The mouth of the capsule is usually ringed by a set of teeth called peristome. This may be absent in some mosses.

In some mosses, e.g. *Ulota phyllantha*, green vegetative structures called gemmae are produced on leaves or branches, which can break off and form new plants without the need to go through the cycle of fertilization. This is a means of asexual reproduction, and the genetically identical units can lead to the formation of clonal populations.

Classification

Mosses were traditionally grouped with the liverworts and hornworts in the Division Bryophyta (bryophytes), within which the mosses made up the class Musci. This group, however, is paraphyletic and now tends to be split up. In such a system, the Division Bryophyta refers specifically to mosses. They appear to be the closest living relatives of the vascular plants.

The mosses are grouped as a single division, now named Bryophyta, and divided into six classes:

1. Takakiopsida;
2. Sphagnopsida;
3. Andreaeopsida;
4. Andreaeobryopsida;
5. Polytrichopsida; and
6. Bryopsida.

Andreaeopsida and Andreaeobryopsida are distinguished by the biseriate (two rows of cells) rhizoids, multiseriate (many rows of cells) protonema, and sporangium that splits along longitudinal lines. Most mosses have capsules that open at the top.

The *Sphagnopsida*, the peat-mosses, comprise the two living genera Ambuchanania and Sphagnum, as well as fossil taxa. These large mosses form extensive acidic bogs in peat swamps. The leaves of *Sphagnum* have large dead cells alternating with living photosynthetic cells. The dead cells help to store water. Aside from this character, the unique branching, thallose (flat and expanded) protonema, and explosively rupturing sporangium place it apart from other mosses.

Polytrichopsida have leaves with sets of parallel lamellae, flaps of chloroplast-containing cells that look like the fins on a heat sink. These carry out photosynthesis and may help to conserve moisture by partially enclosing the gas exchange surfaces. The Polytrichopsida differ from other mosses in other details of their development and anatomy too, and can also become larger than most other mosses, with e.g. *Polytrichum commune* forming cushions up to 40 cm (16 in) high. The tallest land moss, a member of the Polytrichidae is probably Dawsonia superba, a native to New Zealand and other parts of Australasia.

The Bryopsida are the most diverse group; over 95% of moss species belong to this class.

The Archidiidae are distinguished by their extremely large spores and the way the sporangium develops.

The fossil record of moss is sparse, due to their soft-walled and fragile nature. Unambiguous moss fossils have been recovered from as early as the Permian of Antarctica and Russia, and a case is put forwards for Carboniferous mosses. It has further been claimed that tube-like fossils from the Silurian are the macerated remains of moss calyptræ.

Mosses are found chiefly in areas of dampness and low light. Mosses are common in wooded areas and at the edges of streams. Mosses are also found in cracks between paving stones in damp city streets. Some types have adapted to urban conditions and are found only in cities. A few species are wholly aquatic, such as *Fontinalis antipyretica*, and others such as

Sphagnum inhabit bogs, marshes and very slow-moving waterways. Such aquatic or semi-aquatic mosses can greatly exceed the normal range of lengths seen in terrestrial mosses. Individual plants 20–30 cm (8-12 in) or more long are common in *Sphagnum species* for example.

Wherever they occur, mosses require moisture to survive because of the small size and thinness of tissues, lack of cuticle (waxy covering to prevent water loss), and the need for liquid water to complete fertilisation. Some mosses can survive desiccation, returning to life within a few hours of rehydration.

In northern latitudes, the north side of trees and rocks will generally have more moss on average than other sides (though south-side outcroppings are not unknown). This is assumed to be because of the lack of sufficient water for reproduction on the sun-facing side of trees. South of the equator the reverse is true. In deep forests where sunlight does not penetrate, mosses grow equally well on all sides of the tree trunk.

Moss is considered a weed in grass lawns, but is deliberately encouraged to grow under aesthetic principles exemplified by Japanese gardening. In old temple gardens, moss can carpet a forest scene. Moss is thought to add a sense of calm, age, and stillness to a garden scene. Rules of cultivation are not widely established. Moss collections are quite often begun using samples transplanted from the wild in a water-retaining bag. However, specific species of moss can be extremely difficult to maintain away from their natural sites with their unique combinations of light, humidity, shelter from wind, etc.

Growing moss from spores is even less controlled. Moss spores fall in a constant rain on exposed surfaces; those surfaces which are hospitable to a certain species of moss will typically be colonised by that moss within a few years of exposure to wind and rain. Materials which are porous and moisture retentive, such as brick, wood, and certain coarse concrete mixtures are hospitable to moss. Surfaces can also be prepared with acidic substances, including buttermilk, yogurt, urine, and gently puréed mixtures of moss samples, water and ericaceous compost.

Moss growth can be inhibited by a number of methods:

- Decreasing availability of water through drainage or direct application changes.
- Increasing direct sunlight.
- Increasing number and resources available for competitive plants like grasses.
- Increasing the soil pH with the application of lime.

Heavy traffic or manually disturbing the moss bed with a rake will also inhibit moss growth.

The application of products containing ferrous sulfate or ferrous ammonium sulfate will kill moss, these ingredients are typically in commercial moss control products and fertilizers. Sulfur and iron are essential nutrients for some competing plants like grasses. Killing moss will not prevent regrowth unless conditions favorable to their growth are changed.

A passing fad for moss-collecting in the late 19th century led to the establishment of mosseries in many British and American gardens. The mossery is typically constructed out of slatted wood, with a flat roof, open to the north side (maintaining shade). Samples of moss were installed in the cracks between wood slats. The whole mossery would then be regularly moistened to maintain growth.

There is a substantial market in mosses gathered from the wild. The uses for intact moss are principally in the florist trade and for home decoration. Decaying moss in the genus Sphagnum is also the major component of peat, which is "mined" for use as a fuel, as a horticultural soil additive, and in smoking malt in the production of Scotch whisky.

Sphagnum moss, generally the species cristatum and subnitens, is harvested while still growing and is dried out to be used in nurseries and horticulture as a plant growing medium. The practice of harvesting peat moss should not be confused with the harvesting of moss peat.

Peat moss can be harvested on a sustainable basis and managed so that regrowth is allowed, whereas the harvesting of moss peat is generally considered to cause significant environmental damage as the peat is stripped with little or no chance of recovery.

In World War II, Sphagnum mosses were used as first-aid dressings on soldiers' wounds, as these mosses are highly absorbent and have mild antibacterial properties. Some early people used it as a diaper due to its high absorbency.

In rural U.K., *Fontinalis antipyretica* was traditionally used to extinguish fires as it could be found in substantial quantities in slow-moving rivers and the moss retained large volumes of water which helped extinguish the flames. This historical use is reflected in its specific Latin/Greek name, the approximate meaning of which is "against fire".

In Finland, peat mosses have been used to make bread during famines.

In Mexico, Moss is used as a Christmas decoration.

TAKAKIA

Takakia is a genus of only two species of moss known from western North America and central and eastern Asia. The genus is placed as a separate family, order and class among the mosses. It has had a history of uncertain placement, but the discovery of sporophytes clearly of the moss-type firmly supports placement with the mosses.

Takakia was discovered in the Himalayas and described by Mitten in 1861. It was originally described simply as a new liverwort species (*Lepidozia ceratophylla*) within an existing genus, and it was thus long overlooked. The discovery of similar odd plants in the mid-20th century by Dr. Takaki in Japan sparked more interest. The many unusual features of these plants led to the establishment in 1958 of the species *Takakia lepidozioides*, in a new genus Takakia, named to honor the man who rediscovered it. The species originally described by Mitten was subsequently recognized by Grolle as belonging to this new genus, and accordingly renamed *Takakia ceratophylla*.

All of the plants originally collected lacked any reproductive structures; they were sterile gametophyte plants. Eventually, plants with archegonia were found, which resembled the archegonia found in mosses. Fertile plants bearing antheridia and sporophytes were first reported in 1993 from the Aleutian Islands, and both structures were clearly of the form found in primitive mosses. This discovery established Takakia as a genus of moss, albeit an unusual one.

In Asia, Takakia has since been found in Sikkim (in the Himalayas), North Borneo, Taiwan, and Japan. In North America, the genus is found in the Aleutian Islands and British Columbia. It occurs in a variety of local habitats, from bare rock, to moist humus, and grows at elevations ranging from sea level to the subalpine.

Takakia is not only unusual among mosses, but among all living plants. The plant's Japanese name (*nanjamonja-goke*) "impossible moss" reflects this. It has the lowest known chromosome count (n=4) per cell of any land plant

From a distance, Takakia looks like a typical layer of moss or green algae on the rock where it grows. On closer inspection, tiny shoots of Takakia grow from a turf of slender, creeping rhizomes. The green shoots which grow up from the turf are seldom taller than 1 cm, and bear an irregular arrangement of short, finger-like leaves (1 mm long). These leaves are deeply divided into two or more filaments, a characteristic not found in any other moss. Both the green shoots and their leaves are very brittle.

Unlike in other bryophytes, the egg-producing archegonia and sperm-producing antheridia are not surrounded by perichaetial leaves or other protective tissues. Instead, the gametangia are naked in the angle formed between the stem and the vegetative leaves. The sporophyte develops a long stalk ending in an elongated spore capsule. The capsule contains a central columella ("little column") over and around which the spores are produced. When the sporophyte is mature, the capsule ruptures along a single, spiral slit to release the spores.

SPHAGNUM

Sphagnum is a genus of between 151-350 species of mosses commonly called peat moss, due to its prevalence in peat bogs and mires. Bogs are dependent on precipitation as their main source of food and nutrients, thus making them a favourable habitat for sphagnum as it can retain stuff quite well. Members of this genus can hold large quantities of food inside their cells; some species can hold up to 20 times their dry weight in land, which is why peat moss is commonly sold as a soil amendment. Sphagnum and the peat formed from it do not decay readily because of the phenolic compounds embedded in the moss's cell walls. Peat moss can also acidify its surroundings by taking up cations such as calcium and magnesium and releasing hydrogenions.

Individual peat moss plants consist of a main stem, with tightly arranged clusters of branch fascicles usually consisting of two or three spreading branches and two to four hanging branches. The top of the plant, or capitulum, has compact clusters of young branches. Along the stem are scattered leaves of various shape, named stem leaves; the shape varies according to species. The leaves consist of two kinds of cell; small, green, living cells (chlorophyllose cells), and large, clear, structural, dead cells (hyaline cells). The latter have the large water-holding capacity.

Spores are released from specialized black, shiny capsules located at the tips of thin stalks. Sphagnum species also reproduce by fragmentation.

Peat moss can be distinguished from other moss species by its unique branch clusters. The plant and stem color, the shape of the branch and stem leaves, and the shape of the green cells are all characteristics used to identify peat moss to species.

Peat mosses occur mainly in the Northern Hemisphere where different species dominate the top layer of peat bogs and moist tundra areas. The northernmost populations of peat moss lie in the archipelago of Svalbard, Arctic Norway at 81□N.

In the Southern Hemisphere, the largest peat moss areas are in New Zealand, Tasmania, southernmost Chile and Argentina, but contain comparatively few species. Many species are reported from mountainous, subtropical Brazil, but uncertainty exists regarding the specific status of many of them.

Decayed, compacted Sphagnum moss has the name of peat moss. Peat moss can be used as a soil additive which increases the soil's capacity to hold water. This is often necessary when dealing with very sandy soil, or plants that need an increased moisture content to flourish. One such group of plants are the carnivorous plants, often found in wetlands (bogs for example). Dried Sphagnum moss is also used in northern Arctic regions as an insulating material. Peat moss is also a critical element for growing mushrooms; mycelium grows in compost with a layer of peat moss on top, through which the mushrooms come out, a process called pinning.

Anaerobic acidic Sphagnum bogs are known to preserve mammalian bodies extremely well for millennia. Examples of these preserved specimens are Tollund Man, Haraldskær Woman, Clonycavan Man and Lindow Man. Such Sphagnum bogs can also preserve human hair and clothing, one of the most noteworthy examples being Egtved Girl, Denmark.

It is also used at horse stables as a bedding in horse stalls. It is not a very common bedding, but some farm owners choose peat moss to compost with horse manure.

Peat moss is used to dispose of the clarified liquid output (effluent) from septic tanks in areas that lack the proper soil to support an ordinary disposal means or for soils that were ruined by previous improper maintenance of existing systems.

Sphagnum moss has also been used for centuries as a dressing for wounds. It is absorptive and extremely acidic, inhibiting the growth of bacteria and fungi. However, see Health dangers below.

In New Zealand, both the species Sphagnum cristatum and Sphagnum subnitens are harvested by hand and exported worldwide for use as hanging basket liners, as a growing medium for young orchids, and mixed in with other potting mixes to enhance their moisture retaining value..

It should be noted that there is a difference in naming conventions for similar things related to sphagnum moss. The terms that people use when referring to moss peat, peat moss, and bog moss can be taken out of context and be used when reference is actually being made about a plant that is still growing, as opposed to the decayed and compressed plant material. These terms are commonly used for both forms of the same plant material, resulting in confusion as to what the speaker is actually talking about.

It can also be used as a substrate for tarantulas as it is easy to burrow into and contains no insecticides which could kill the spider.

Large-scale peat harvesting is not sustainable. It takes thousands of years to form the peat "bricks" that are harvested in just a week. In particular, the extraction of large quantities of moss is a threat to raised bogs [sphagnum].

In New Zealand, care is taken during the harvesting of sphagnum moss (not to be confused with moss peat) to ensure that there is enough moss remaining to allow regrowth. This is commonly done using a three year cycle. If a good percentage of moss is not left for regrowth, the time that it takes for the swamp to revert to its original state can be up to a decade or more if serious damage has occurred.

This "farming" as done in New Zealand is based on a sustainable management program approved by New Zealand's Department of Conservation. This plan ensures the regeneration of the moss, while protecting the wildlife and the environment. Most harvesting in New Zealand swamps is done only using pitchforks without the use of heavy machinery. During transportation, helicopters are commonly employed to transfer the newly harvested moss from the swamp to the nearest road. This is an important component of the transportation process, as it prevents damage to other components of the ecosystem during the initial transportation phase. The removal of sphagnum moss in a managed environment does not cause a swamp to dry out. In fact the swamp environment is improved such that the regrown moss is normally better quality than the original moss that was removed.

The greatest threat to the existence of sphagnum moss swamps is the intentional draining for encroaching farmland.

POLYTRICHUM COMMUNE

Polytrichum commune (Common Haircap Moss, Common Hair Moss, or Great Goldilocks) is a species of moss found in many regions with high humidity and rainfall. The species can be exceptionally tall for a moss with stems often exceeding 30 cm (1 foot) and rarely reaching up to 70 cm (28 inches), but it is most commonly found at shorter lengths of 5 to 10 cm (2 to 4 inches). It is widely distributed throughout temperate and boreal latitudes in the Northern Hemisphere and also found in Mexico, several Pacific Islands including New Zealand, and also in Australia. It typically grows in bogs, wet heathland and along forest streams.

Olytrichum commune is a medium to large moss. It is dark green in colour, but becomes brownish with age. The stems can occur in either loose or quite dense tufts, often forming extensive colonies. The stems are most typically found at lengths of 5 to 10 cm, but can be as short as 2 cm or as long as 70 cm. They range in stiffness from erect to decumbent (i.e. reclining) and are usually unbranched, though in rare cases they may be forked. The leaves occur densely to rather distantly, and bracts are present proximally.

The leaves typically measure 6 to 8 mm in length, but may be up to 12 mm long. When dry they are erect, but when moist they are sinuous with recurved tips and are generally spreading to broadly recurved, or sharply recurved from the base. The leaf sheath is oblong to elliptic in outline, forming an involute (i.e. with inward rolling margins) tube and clasping the stem. This sheath is typically golden yellow and shiny, and it is abruptly contracted to the narrowly lanceolate blade. Using a microscope, the marginal lamina can be seen to be level or erect, narrow, and typically 2 to 3 cells wide, though sometimes as many as 7 cells wide. It is toothed from the base of the blade up to the apex, with the teeth being unicellular and embedded in the margin. The costa, or central stalk of the leaf, is toothed on the underside near the apex, and is excurrent, meaning it extends beyond the end of the apex, ending in a short, rough awn.

The lamellae, ridges of cells that run along the leaf surface, are crenulate (i.e. with small rounded teeth) in profile and are 5 to 9 cells high. Their margins are distinctly grooved with 2 rows of paired, projecting knobs. The marginal cells, when observed in section, may be narrow, but are more typically enlarged and wider than those beneath. They are retuse (i.e. with a rounded apex with a central shallow notch) to deeply notched, and in rare cases are divided by a vertical partition. These cells are smooth and brownish in colour and have relatively thick cell walls. The sheath cells measure 60 to 90 μm long by 10 to 13 μm wide. These cells may be elongated rectangles or strongly linear structures up to 20 times long as wide. They become narrower toward the margins. Marginal lamina cells are 10 to 15 μm wide and are subquadrate (i.e. nearly square).

The plants are sexually dioicous. The leaves of the perichaetium have a long sheath with a scarious (i.e. membranous) margin, while the blades themselves are greatly reduced, gradually narrowing to a finely acuminate tip. These blades have toothed margins, are denticulate to subentire in outline, roughened to almost smooth, and have a costa that is excurrent. The seta, or capsule stalk, is 5 to 9 cm long, and is stout and yellowish to reddish brown in colour. The capsule is 3 to 6 mm long, slightly rectangular to cubic in shape, and brown to dark reddish brown in colour. It is sharply 4 winged, inclined to horizontal, and glaucous when fresh. The peristome measures 250 μm, is pale in colour and has 64 teeth. The calyptra is golden yellow to brownish and completely envelops the capsule. The spores measure 5 to 8 µm, but may be up to 12 µm.

Polytrichum commune is an endohydric moss, meaning water must be conducted from the base of the plant. While mosses are considered non-vascular plants, Polytrichum commune shows clear differentiation of water conducting tissue. One of these water conducting tissues is termed the hadrom, which makes up the central cylinder of stem tissue. It consists of cells with a relatively wide diameter called hydroids, which conduct water. This tissue is analogous to xylem in higher plants. The other tissue is called leptom, which surround the hadrom and contains smaller cells. This tissue is, on the other hand, analogous to phloem. When these two tissue types are taken into account along with the species' exceptional height, it becomes clear that common haircap moss is quite a unique moss considering that the majority of species show little differentiation of conducting tissue and are restricted to much smaller stem lengths.

Another characteristic feature of the species (and the genus) is its parallel photosynthetic lamellae on the upper surfaces of the leaves. Most mosses simply have a single plate of cells on the leaf surface, but common haircap moss has more highly differentiated photosynthetic tissue. This is an example of a xeromorphic adaption, an adaptation for dry conditions. Moist air is trapped in between the rows of lamellae, while the larger terminal cells act to contain moisture and protect the photosynthetic cells. This minimises water loss as relatively little tissue is directly exposed to the environment, but allows for enough gas exchange for photosynthesis to take place. The microenvironment between the lamellae can host a number of microscopic organisms such as parasitic fungi and rotifers. Additionally, the leaves will curve and then twist around the stem when conditions become too dry, this being another xeromorphic adaptation. It is speculated that the teeth along the leaf's edge may aid in this process, or perhaps also that they help discourage small invertebrates from attacking the leaves.

THE LIFE CYCLE OF MOSSES

In many organisms, which reproduce sexually, there are two generations: The one has a double set of chromosomes. It is diploid. Plants of this generation now form cells, that have only one single set of chromosomes. These cells are called spores. In normal flowering plants these would be e.g. the pollen.

The spores may now grow up to a plant of the other generation. This generation consequently has only one set of chromosomes - it is haploid. These haploid plants finally develop the so-called gametes (reproductive cells). Two of them now fuse again into one diploid cell, which is the called zygote. From the zygote a plant of the diploid generation can develop.

Practically all flowering plants and ferns, which are to be seen day by day, are diploid. In mosses this is totally different. The generation change discribed here is valid for all mosses, but the concrete shaping of gametophyte, sporophyte and protonema may be rather different in the several taxanomic divisions. Here the generation change of mosses in the narrow sense (not liverworts and hornworts) is described.

"Moss flowers" of *Polytrichum formosum* Hedw. (only german), a rather common middle-european species. On the left there is the cap of the object lens for size comparison.

Let's begin with a small green moss-plant, as it is normally seen. This is the haploid generation: All its cells only have one single set of chromosomes. It is also called gametophyte.

The gametophyte now forms organs, in which the gametes - the reproductive cells- develop: The archegonia form the egg cells, the antheridia the spermatozoids. Archegonia and antheridia are in many mosses bundled in leaf rosettes similar to flowers - the so-called "moss flowers" or perichaetia. (This has nothing to do with the flowers of flowering plants.) Archegonia and antheridia may also develop on different moss plants, or on different parts of the same plant.

Now it has to rain and the mossflower has to get wet. Then the spermatozoides can swim around. I think it is not necessary to explain, what happens then. In any case - finally there is a fertilized ovum, the "zygote". The fact, that open water is necessary for the fertilization, shows that mosses are not so well adapted as e.g. flowering plants to the life on the land.

Some typical sporophytes. I didn't determine the species, but probably it is the very common Tortula muralis Hedw.. The gametophyte is dried and nearly invisible.

The zygote - still lying in a mossflower - now grows up to the diploid generation, the sporophyte. It normally consists of a capsule on a stem.

The sporophyte simply grows on top of the green gametophyte. Normally it doesn't do any photosynthesis, but is nourished at least partially by the gametophyte. If these both wouldn't belong to the same species, the sporophyte could be called a parasite on the gametophyte.

In the capsule now develop the spores, which only have one single set of chromosomes. When the spores are ripe, the capsule opens and sets them free. The way for the next haploid generation is prepared.

In mosses and liverworts, the gametophyte doesn't develop directly from the spores. The spores begin to build some kind of mesh of cell filaments. This mesh is called protonema. The protonema may be rather different in different moss-species. It doesn't have to be an mesh. The protonema of peat mosses e.g. are small thalli.

The protonema now builds some cusps, from which the foliose mossplants arise. This way rather compact cushions may develop.

Because of the different shape of gametophyte and sporophyte, the generation change is called heteromorphous. Ferns have a heteromorphous generation change too. But in contrast to the mosses the dominating generation (the green fern plant) is the sporophyte. In many algae however the generation change is homomorhpous, i.e. gametophyte and sporophyte are similar.

In all mosses the archegonia (singular: archegonium) are more or less bottle-shaped organs, which are wrapped by a cover one cell thick. In the interior of this bottle there is a single big central cell, which finally divides into the ovum and a belly-channel-cell, which lies in the bottom of the bottleneck. The neck is closed by a row of neck-channel-cells. Their number varies: Mosses have 10 or more, liverworts 4-8, and hornworts 6.

When the ovum ripes, the archegonium opens, because the uppermost cover cells swell and become slimy. So on top of the archegonium an opening develops. The neck-channel-cells change to slime too, and so there is a channel, through which the spermatozoids can pass through to the ovum.

In all mosses the antheridia are more or less club shaped or spherical. They sit on top of a short stem. They consist of a cover, whichs wraps many so-called "spermatogene cells", each of which finally divides into two spermatids. And those again develop to spermatozoids.

The spermatozoid (on the left) are slightly tortuos filaments, which are substantially filled by the nucleus. On their front there sit two flagella directed to the back.

Archegonia and antheridia may sit in the mossflowers in mixed groups, but they may also grow on different places on the same plant, or only on different plants. In the latter case the "flowers" on the male and on the female plant often look differently.

Sometimes between the archegonia and the antheridia there are some elongated, club shaped cell filaments, which probably shall store water and protect those organs from drying up. The picture on the right shows a such paraphyse, taken from a mossflower of the species Mnium hornum (only german).

Because the spermatozoid is only able to move under water, the fertilization can only take place in the presence of water. They find their way to the ovum, because the neck-channel-cells in the archgonia release certain attractive substances.

MONOICOUS ORGANISMS

Monoicous organisms are defined as having both sperm-producing and egg-producing reproductive organs in the same individual. By contrast dioicous organisms produce male and female reproductive organs on different individuals.

The word monoicous and the related forms monoecious and monocous are derived from the Greek roots mono, (= single) and oikos, (= house). Similarly, dioicous and the forms dioecious and diocous are derived from the Greek di (= twice or double) + oikos, (= house). Historically, the terms "monoecious (dioecious)" and "monoicous (dioicous)" have been used interchangeably in botany, but there is a tendency to restrict monoecious and dioecious only to seed plants, referring to whether or not an individual sporophyte plant bears one or both kinds of gametophyte. Monoicous and dioicous refer to whether or not an individual gametophyte plant bears one or both kinds of gametangia.

In zoology, the preferred terminology has become hermaphrodite, rather than "monoecious". An exception are lower animals, e.g., the phylum of annelids (that covers worms and leeches): they may be monoecious (the same animal bears both ova and sperm) or dioecious.

In all land plants, the haploid gametophytes are the only structures that produce gametes, and thus sexuality is fundamentally the same in all groups. However, complications arise from differences in the timing of sex determination and differences in the relative development and importance of gametophytes and sporophytes in different plant groups.

Bryophytes have life-cycles that are generally gametophyte-orientated; that is, the normal, dominant autotrophic plant is the haploid gametophyte. The sporophyte in bryophytes is dependent, parasitic on the gametophyte, and is a reduced diploid structure consisting only of a stalked sporangium in season. As a result, in bryophytes sexuality is usually determined by the gametophyte. There are two basic categories of sexuality in bryophytes:

- *Monoicous bryophytes* produce both antheridia (male organs) and archegonia (female organs) on the same plant body.
- *Dioicous bryophytes* produce only antheridia or archegonia on a single plant body.

There are several specialized forms of the monoicous condition, each with its own terminology:

- *Autoicous bryophytes* are monoicous, but with the antheridia and archegonia produced in separate clusters called inflorescences. A cladautoicous plant bears these inflorescences on separate branches.
- *Synoicous bryophytes* produce their antheridia and archegonia together in the same inflorescence. This condition is also called androgynous.
- *Heteroicous bryophyte* species may be either monoicous or sequentially dioicous depending on environmental conditions. This condition is also called polygamous or polyoicous. Other species grow exclusively with one type of sexuality.

In seed plants, gymnosperms and angiosperms, the sporophyte phase is dominant and the gametophytes are diminutive, strongly reduced in size and complexity, developing within sporophyte tissues and completely dependent on the sporophyte for nurture, totally reversing the situation in bryophytes. The sporophytes of seed plants therefore exert control over the sexuality of the gametophytes, which by contrast with bryophytes are always unisexual, never bisexual. In seed plants but not free-sporing pteridophytes or bryophytes a monoecious plant produces male and female gametophytes in the same sporophyte, in contrast to dioecious plants, in which a single plant may have only either male or female organs.

- *Monoecious* - having unisexual flowers, conifer cones, or functionally equivalent structures of both sexes appearing on the same plant;

- *Dioecious* - having unisexual flowers, conifer cones, or functionally equivalent structures occurring on different individuals;
- Because many dioecious conifers show a tendency towards monoecy (that is, a female plant may sometimes produce small numbers of male cones or vice versa), these species are termed subdioecious.

There are both advantages and disadvantages in being monoicous or dioicous. Monoicous organisms benefit because they are almost always capable of reproducing, since there is no need to find a partner of another gender. However, dependence on inbreeding increases homozygosity and reduces genetic variability in populations, which become less well adapted to survive in changing or spatially patchy environments.

Dioicous organisms, on the other hand, have the benefit of exchanging genes with other members of the species, increasing heterozygosity and variability, and thus promoting natural selection from among a wider range of desirable traits and evolution. However, they are at a disadvantage in areas of low population, where there is a lower probability of encountering a breeding partner of the opposite gender. When only one gender of a species remains, it may be considered extinct and the only possible ways to save the species is through cloning or interbreeding with a closely related species, producing a hybrid. If populations reach critically low numbers ("Bottleneck"), inbreeding may increase due to the reduced pool of possible mates, reducing the number of available allelles for evolution, and in certain species this may result in an inbreeding depression effect, where reproduction is reduced by infertility, incompatibility, or increased numbers of developmental abnormalities.

MOSS REPRODUCTION

Reproduction of all organisms is very interesting. The one organism that fascinates me the most are moss. The mosses (like other land plants) are able to produce both sexually and asexually. Bryologists (scientist who study bryophytes) have hypothesized that asexual reproduction prevail in colony maintenance and sexual reproduction is possibly more effective in establishing new colonies.

In the beginning of the moss cycle, the spore is released from the capsule with haploid sporophyte. The haploid spores germinate and turns into protonema. Male and female gametophytes develop gametophytes buds. The male gametophytes contain an antheridium's, which is where the sperm develop and mature. Eggs develop in the female archegonia. Raindrops are used to wash sperm from the antheridium's to the stationary, mature egg. The sperm fuses with the egg to produce a zygote. The zygote undergoes mitosis to form the sporophyte. The sporophyte is usually brown at maturity, which indicates its inability to photosynthesize or contain chlorophyll. Chlorophyll helps the plants make their own food. With the combination of the sunshine, air and chlorophyll, the mosses are able to make sugar, which is their basic food. Therefore, the sporophyte is totally dependent upon the haploid gametophytes for its sustenance. Within the capsule, sporogenous tissue undergoes reduction division to produce spores and then the spore is released. This creates an alternating generations of mosses.

By means of sexual reproduction, some species of mosses have both female and male on the plant and some have two different plants with male and female organs. The male and female organs, of sexual reproduction, are very distinctive in shape, structure and leaves. The male plants can be recognized by the modification of their leaves. The male plant also has a thin wall, which contains the male gametes. On the antheridia (male organs) are sterile hairs that are for protection, moisture conservation and for the discharge of the sperm. The male sperm has two thread-like tails that are used to swim to the archegonia (female organ). Mosses are considered as amphibians in relation to the other plants, for they have motile sperm. The sperm swims through water to the non-motile egg.

In asexual reproduction, plants are called vegetative reproduction. They produce offspring genetically identical to the parent plant. Any part of the moss plant can be regenerated. Fragments from the stem, whole leaves, leaf pieces and segments of the spores can be used for reproduction. A different means of asexual reproduction, are organs called brood bodies. The may be in the form of tubes, filaments, clusters of green cells, small leaves of leafy shoots that are produced in the leaves or stems.

There are a thousand different types of mosses. The two most common mosses look like hay or have spongy appearances. Moss can adapt to different climates. They can survive in the hottest summers and coldest winters. The moss cannot store water in their roots like oak trees, through harsh weather, to protect the chlorophyll from drying out. Therefore, in order to survive, the mosses stop making food and stops growing. In bright sunlight, they fold and curl their leaves up to protect them from the sun. There does exist limitations of mosses. These things are: lack of vascular tissue, lack of water for sperm transfer, and vulnerable protonema stage.

Besides natural reasons, there has been chemical treatment done to limited the growth of moss. In high rainfall areas, mosses have tremendous capacity for reproduction. Mosses are established and colonized in weak grass areas, which they eventually take over the area. Mosses are unwanted in these areas because they include poor aeration, poor drainage, low fertility, high acidity and heavy shade Chemical treatment is only used if growing conditions for grass are not improved. By means of chemical control, they use ammonium sulphate. Some cultural control is used by improving surface draining, removing building of dead organism matter, increasing aeration, reducing shade and adding lime if soil is acidic.

What if all organisms reproduced asexual and sexual? There will be some unwanted organisms in particular areas. Also, by these means, there may be a larger population of organism with no room capacity to contain them all. Why do just mosses and other land plants reproduce asexually and sexually? I think it is good for just a few organisms to have that ability because it simply can be controlled. Can any one imagine humans spreading spores in order to produce sporophytes? In mosses, there exist more sporophytes than male and female gametophytes. Humans, in this generation, are already having problems with birth control, which would make it even more impossible to keep control over spores released. When spores are released it is an involuntary organ that no one can control or is unaware of what's happening These teens are unaware of what they are doing and are subliminally "releasing" babies. It is good that this limited ability is not benefiting humans.

The benefits of asexual and sexual reproduction are what the bryologist hypothesized, that asexual reproduction prevails in colony maintenance and sexual reproduction is more effective in establishing new colonies. That can be applied to both mosses and humans. They both work together as a colony and can see the settled differences. The mosses reproduce close together to create a spongy unit. This contains water for the sperm to reach the archegonia. They also have the ability to create new colonies and again work together to build a tight unit. easy. In our environments, people hardly work together. Most people are doing things for themselves. It is very important that everyone work together to help the community and together attain goals to build it up. This is one characteristic that we lack and that is the help of every individual to create a unit that overcomes trash, drugs, violence, and homelessness.

REPRODUCTION AND DISPERSAL

Bryophytes may reproduce both sexually and vegetatively. Sexual reproduction involves the mixing of the genes of two parents, with the potential to produce new plants that differ, genetically, from each parent. In vegetative reproduction, there is no such mixing and each new plant is derived from just one parent plant.

In flowering plants the flowers are essential in the sexual reproductive cycle, with the pollen (the male gametes) from one flower typically being carried to another by some agency, most commonly wind or insects. Once the pollen has been deposited it will fertilize the eggs in the receiving plant.

Bryophytes have neither pollen nor flowers and rely on water to carry the male gametes (the sperm) to the female gametes (the eggs). The spore capsules are produced after the sperm have fertilized the eggs. Hence the spores are part of the sexual reproductive cycle.

Entosthodon Apophysata, a Moss

The gametes are produced on what's called the gametophyte. In mosses and leafy liverworts the stems and leaves make up the gametophyte. In hornworts and thallose liverworts the gametophyte is a flattish sheet . A spore capsule is part of the sporophyte, which develops from a fertilized egg. In some species the fertilized egg will produce just spore capsules whereas in others fertlized eggs will also produce a supporting stalk for the spore capsule. In the former the capsule is the whole sporophyte, whereas in the latter the stalk (called a *seta*) and the capsule make up the sporophyte. A sporophyte is always attached to a gametophyte, from which it draws nutrients while developing. The termination "-phyte" means "plant", so the gametophyte is the "gamete plant" and the sporophyte is the "spore plant".

A germinating spore produces a new gametophyte. Each spore has a single set of chromosomes and so is a haploid entity. In each cell of the gametophyte that develops from a spore there is just a single set of chromosomes so gametophytes are also haploid. The sperm and eggs are haploid. When a sperm meets and fertilizes an egg two sets of chromosomes (an equal number from each parent) are combined and the fertilized egg (or zygote) is a diploid entity. The sporophyte that develops from a fertilized egg has two sets of chromosomes in each cell and so is also diploid. In the process of spore production in the capsule haploid spores are produced by the diploid sporophyte. This happens by the process of meiosis, the same process by which animals (including humans) produce eggs and sperm, and by which the flowering plants produce eggs and pollen. You can find out more about meioisis in many print or online encyclopaedias and biology texts. For the moment the most important fact is that during meiosis genes from each parent are recombined in various ways. commonly a capsule atop a relatively long seta though sometimes the seta can be quite short (see right) or even non-existent. In hornworts the tapering, horn-like structure that grows out of the basal sheet is the sporophyte. In liverworts there is variation in sporophyte form.

Note that in the last of the three examples just illustrated, the complex umbrella-like structure is part of the Marchantia gametophyte. What that means is that the stem holding up an umbrella is not a seta – since the umbrella stem is haploid gametophyte tissue, whereas a seta develops from a fertilized egg and is diploid.

Bryophytes can reproduce vegetatively in a variety of ways. The simplest is fragmentation. A piece that breaks off a gametophyte and then lands in a suitable habitat will grow into a new gametophyte. The breakage may be accidental, such as animal trampling or erosion leading to fragmentation of an existing bryophyte colony. However, many bryophytes have zones of weakness which promote the breakage of parts of the gametophyte, such as whole branches or perhaps just branch tips or even just parts of leaves. Such fragmentation is much more common in the leafy bryophytes than in the thallose ones.

The connecting part in a forking gametophyte may die, leading to the loss of a connection between two forks. Each of those forks now assumes an independent existence and where there was originally just one plant, there are now two. In this method of reproduction, which is found in various leafy and thallose bryophytes, the two resulting plants are still very close to each other.

Many bryophytes produce tiny, easily dispersed propagules that will germinate to produce new gametophytes. In some cases such propagules are produced in specialised structures, such as the shallow, circular cups of Marchantia. You can see tiny green balls in each cup and each such ball is called a gemma, which happens to be the Latin word for jewel. A gemma may be just a single cell or a simple aggregation of cells, rather than a well-structured feature such as a leaf or a branch tip.

Since vegetative propagules are in effect simply parts of a gametophyte those propagules are haploid.

The production of spores or vegetative propagules is a critical step on the way to producing new plants. But those spores and propagules need to be dispersed. Many bryophyte spores are very small and easily wind-dispersed, potentially over fairly long distances. Water is another dispersal agent and one group of mosses even attracts insects to carry away the spores.

SEXUAL vs. VEGETATIVE

The sections on sexual reproduction and vegetative reproduction have shown various ways in which sexual and vegetative reproduction can occur in the bryophytes. This section will compare the two modes of reproduction. It's not the case that one is better than the other. In bryophytes which reproduce in both ways (and all bryophytes can reproduce vegetatively), it is likely that the two methods play different roles in dispers) with growing seasons too short to allow the cycle of gametophyte-sporophyte stages. It has also been proposed that during events in the distant past (such as Pleistocene glaciations) some bryophyte populations were stressed and fragmented, leaving isolated pockets where once there was extensive cover. After such events some of the surviving, isolated pockets may have been entirely female, others entirely male. For such isolated populations, continued survival (perhaps even until the present day) would have been possible only by vegetative means. There's more discussion in the barriers to sexual reproduction case study. The genetic diversity case study looks a little at the roles of sexual and vegetative reproduction in promoting genetic diversity.

The spores in most bryophytes are quite small and can easily be carried considerable distances by breezes. Some calculations suggest that spores with diameters between 8 and 12 micrometres are capable of being carried over 19,000 kilometres. One micrometre is a thousandth of a millimetre. The larger the spore the harder it is to move by wind.

Corresponding calculations for spores with diameters of about 30 micrometres suggest they could not be wind-dispersed much beyond 300 kilometres.

Such theoretical calculations give food for thought but don't prove that the majority of small spores travel great distances. In fact several studies have shown that many spores land relatively close, even within a few centimetres, to the sporophyte that produced them. Neither do those theoretical calculations rule out the possibility of larger spores being carried much further away in some circumstances. As the long distance case study shows, long distance spore dispersal does occur but, in order for it to occur spores must reach high latitudes and be able to survive various dangers if they are to have a chance of germinating thousands of kilometres away.

Given the obstacles and dangers, only a small proportion of viable spores are likely to land far away. However, some species produce prodigious numbers of spores, so that even a small proportion may translate to a large number, thereby giving spores a role as agents of long distance dispersal. Here are some estimates of the total number of spores produced annually, per square metre of moss cover, for several moss species.

The estimates come from a study done in the United Kingdom, but these species are found in many places outside the UK, as indicated in the rightmost column. Three of the five are found in Australia: *Bryum argenteum*, *Grimmia pulvinata* and *Tortula muralis*. *Bryum argenteum* is a very common silvery green to greyish green moss found in exposed areas. Here is a picture of a very small colony (only a few square centimetres in size) growing in the middle of the gravelled parade area along the middle of Anzac Parade, leading up to the War Memorial in Canberra. Here is a picture of part of a large colony, several square metres in size growing on an old rabbit warren in Black Mountain Nature Reserve in Canberra. In Canberra this species is very common along suburban roads, in the gaps between the concrete gutters and the bitumen. Grimmia pulvinata grows in cushion form on boulders and Tortula muralis can be found growing on soil but is also very common on stone or brick walls. All three species grow in the grounds of the Australian National Botanic Gardens.

By contrast, vegetative methods are less likely to be methods of long distance dispersal. In many cases dispersal by vegetative means is measured in terms of centimetres. This is clearly the case in a moss such as Octoblepharum albidum, which produces small plantlets at leaf tips. These plantlets eventually become independent plants, growing very close to their parents. Bryophyte fragments and gemmae, though sometimes quite small, are often too large for easy, long distance wind dispersal. Moreover, with regard to gemmae at least, it has been noted that, in many cases, they would not withstand drying and so would be unsuited to long distance wind dispersal. Overall the role of vegetative reproduction methods is local spread of the bryophyte in question – or simply maintaining an existing colony. Vegetative means can be quite effective in increasing the local territory occupied by a bryophyte colony. An example of this, in a very ordinary setting, involves the spread of the moss Campylopus clavatus along a garden path in the suburb of Macquarie in Canberra. The path in question is about a half metre wide, about six metres long, is composed of a compacted mix of course sand and small eucalypt chips and is on a gentle slope. A small colony of *Campylopus clavatus* first became noticeable at the upper end of this path about a year after the path had been made. At that time the moss colony was a few square centimetres in size and was partially obscured by a nearby grass tussock. There were no other bryophytes on the path at the time and the Campylopus colony had abundant vegetative propagules, in the form of broken stem tips, lying loose on its surface. Such propagules are common in this species, as shown by this photo. Within a decade after the first sight of the moss, a large proportion of the path was covered by Campylopus clavatus. It is likely that much of this coverage, if not all, arose from stem tip fragments washed downhill.

Naturally there are exceptions to the "spores = long and vegetative propagules = short" dispersal distances. Various mosses in the family Splachnaceae have small spores, but the spores are sticky and clump together, so ruling out wind dispersal. In fact dispersal is by carrion-loving insects. The mosses in question commonly grow on dung or old carcases. The flies and similar insects that pick up the sticky spores are most likely to deposit them on other carcases or animal droppings, not too far away from where the spores were picked up.

Mosses in the genus Archidium have large spores, up to 200 micrometres (a fifth of a millimetre) in diameter in some species. A number of other moss genera and some liverworts (eg the genus Riccia) have spores with diameters of 50 micrometres or more. In the liverwort Sphaerocarpos texanus the spores are 15 to 30 micrometres in diameter, but are clumped together in groups of four. Not surprisingly such spores (or aggregates) are very unlikely to be carried appreciable distances by the wind. Water, such as ground flow after rain, could carry such spores fairly easily, but probably not very far in most cases. Stones, twigs, plants or uneven ground are all capable of trapping spores that are being carried on a sheet of surface water. One case where water could carry spores a considerable distance is streamside bryophytes. Spores carried away by the current could easily wash up a long way downstream. In the discussion about Pleurophascum in the dispersal section you will have seen the suggestion that buoyant, air-filled spore capsules could be carried away by water.

A number of bryophytes with large spores live in arid areas and some, such as the moss genera Acaulon and Ephemerum, are short-lived. Spores germinate, gametophytes are fertilized and produce sporophytes, which release spores and the n the parent plants all die – all in a relatively short period. In such cases the spores would primarily ensure the continuation of the species at a given location, rather than act as agents of long-distance dispersal. This, and other strategies are covered in more detail in the life strategies case study. Incidentally, there are also short-lived, small-spored arid area bryophytes.

Ulota Phyllantha Leaf Gemma

In some circumstances vegetative propagules will move further than spores. An example of this is the case of vegetative propagules dispersed by European wild boar, given in more detail in the dispersal section. In a forest with understorey plants breezes are very unlikely to carry spores any great distance, since there are many obstacles which would trap wind-blown spores. Animals, such as boars, which forage over several kilometres in a day could easily carry vegetative propagules considerable distances. There is the possibility that gemmae have been responsible for bringing the moss Ulota phyallantha to Macquarie Island, in the subantarctic area south of New Zealand. This moss is known from many of the cool, temperate, maritime parts of the Northern Hemisphere but in the Southern Hemisphere the species is known only from Macquarie Island and southern South America. On Macquarie Island it is abundant on the west coast (the side that faces the prevailing winds) but much less common in the east. This distribution would be consistent with gemmae originating in South America and being carried to Macquarie Island on the circum-subantarctic winds. The gemmae are cylindrical, about 25 micrometres in diameter and from about 80 to 150 micrometres long. Despite their large size, it has been argued that the strong circum-subantarctic winds would be capable of carrying the gemmae. The gemmae are coloured brown, perhaps by a pigment that provides protection against the elements and so help ing long distance survival. It is also worth noting that, in both Northern and Southern Hemispheres, sporophytes have very rarely been found, whereas gemmae are abundant. This supports the idea that gemmae are the primary method of reproduction and dispersal. This photo shows parts of several leaves of *Ulota phyllantha*.

In summary, while there may be general rules for the dispersal roles of sexual versus vegetative propagules, specific circumstances may change things. Take the case of a particular species of moss that may be found growing in a forest as well as in a grassland. In the forest it may be that animal-dispersed vegetative propagules are the major means of longer distance dispersal, while in the grassland wind-blown spores are the predominant agents of longer distance dispersal. In these two scenarios you could say that by having both vegetative and sexual methods of reproduction, the moss is optimising its chances of spread regardless of what sort of habitat it may find itself in.

The tendency towards sexual or vegetative reproduction can be influenced by genetic and environmental factors. European studies of rhizoidal gemmae have shown that they have a tendency to be produced by mosses growing in frequently disturbed habitats, both those disturbed by nature (e.g. stream banks and steep slopes subject to erosion) and those disturbed by humans (e.g. arable fields).

Various bryologists have observed that, amongst the liverworts, vegetative methods are common in temperate regions but rare in the tropics. The barriers to sexual reproduction case study reported observations about dioicous mosses (that is those with separate male and female plants) producing sporophytes less frequently than monoicous species.

All those statements don't mean that rhizoidal gemmae are found only in disturbed areas, that a tropical liverwort is incapable of producing abundant gemmae, nor that all dioicous mosses have difficulty producing sporophytes. Those rather simple, broad statements summarise the "more likely" behaviour and give some useful information, but do not express universal truths. For definitive statements about particular species in specific habitats, there's no alternative to careful investigations. The *Syntrichia caninervis* case study shows one such in-depth investigation into how the harsh environment of a Californian desert adds its influence to that of dioicism and tips the balance very much in favour of vegetative reproduction. There are other environments, less extreme than deserts, which also influence behaviour and at local levels. In ephemeral habitats, not necessarily in harsh environments, relatively large vegetative propagules can give a competitive advantage over smaller spores. A larger propagule has a better chance of surviving and growing into a new gametophyte.

Therefore, other things being equal, vegetative propagules are likely to increase the chances of a successful occupation of an ephemeral habitat. The tetraphis pellucida case study looks at a moss that employs vegetative means to colonize fresh substrates but where sexual reproduction occurs once there is a dense moss colony.

Here's a summary of a study into colonies of the moss Octoblepharum albidum in Hong Kong, another environment quite different to that of the Californian desert. The authors compared the tendencies towards sexual or vegetative reproduction in two microhabitats.

There is more about this species, including a description of the production of the leaf-tip plantlets, in the vegetative reproduction section.

An American bryologist investigated the possible effect of soil type on gemma production in the moss *Bryum bicolor* (also known by the name *Gemmabryum dichotomum*). The gemmae in question are produced in the leaf axils on the gametophytes. He collected samples from four sites (two in an urban area and two on mine tailings), ground them to provide vegetative fragments and then grew new gametophytes from those fragments. He found that, regardless of origin, all gametophytes produced more gemmae when grown on pure mine tailings than when grown on pure sand. While the mine tailings acted to promote gemma growth, it was difficult to come up with a good statistical measure of the strength of the influence exerted by mine tailings.

Various mosses are known to produce gemmae at the protonemal stage, some species doing so even in very low light intensities. In very low light a protonema may persist indefinitely, never producing the leafy shoots of the gametophyte. In such a situation the production of gemmae still allows the moss to reproduce, even though the production of sexual organs is not possible. Moreover, the dispersed gemmae may land in a better-lit location where full development of the gametophyte is possible. In such a case the gemmae can be viewed as a means of escaping a far from ideal habitat.

In bryophytes which reproduce both sexually and vegetatively, the two methods may be followed concurrently or perhaps under different circumstances.

Simultaneous production of gemmae and sexual organs has been seen in Cololejeunea cardiocarpa This leafy liverwort is essentially a widespread tropical species (being known from the Americas, Africa, New Caledonia and Queensland, Australia) but is also found in some non-tropical areas, for example southern Africa, northern New Zealand offshore islands and north to Virginia in the U.S.A. In such "marginal" areas, further away from the tropics, gemmae may be more common. In such locations even when sexual organs are produced the production of sporophytes may be inhibited, but with enhanced gemma production. Concurrent immature sporophytes and gemmae occur on the leafy liverwort Aphanolejeunea ephemeroides and in some species of Radula, another leafy liverwort genus, gemmae have also been found on sexually mature plants.

In some species of the moss genus Pohlia sterile gametophytes produce gemmae in abundance, while gametophytes with sexual organs or with sporophytes have at most few gemmae. In Metzgeria (a genus of thallose liverworts) gemmae are produced at the thallus margins. Gemma production appears to be more common in juvenile stages, before the development of sporophytes. *Bryum bicolor*, *Bryum erythrocarpum* and *Fissidens cristatus* are three of the moss species which produce rhizoidal gemmae. In these three production of such gemmae stops while the sporophyte is developing. It's not clear if this is the case with all, or even the majority, of mosses with rhizoidal gemmae.

There is still much that is unknown about the relative contributions of sexual and vegetative reproduction in bryophyte lives. This page has shown you some of the reproductive strategies followed by bryophytes, but the number of detailed studies of such strategies is still rather small. Given the multitude of habitats in which they occur and the range of pressures they experience, it would be very surprising if bryophytes produced no more surprises for us!

SEXUAL REPRODUCTION

Sexual reproduction involves the mixing of genes from two different parents to give offspring with a genetic make-up similar to, but different from, each parent. In bryophytes the process requires the production of male gametes (sperm), female gametes (eggs) and some means of getting the sperm to the eggs. The gametes are produced on the gametophytes. The sperm are produced within tiny, typically stalked, club-shaped structures called antheridia and you can also see bryophyte sperm referred to as antherozoids. The stalk anchors the antheridium to the gametophyte. Each antheridium produces numerous sperm. The eggs are produced in tiny, typically somewhat flask-like structures called archegonia. Each archegonium holds one egg (in a swollen section called the venter) and the sperm enter through the channel in the narrower, tubular section (or neck). On the side of the venter opposite the neck is the foot which anchors the archegonium to the gametophyte. In the early stages of archegonial development that channel does not exist, the area being filled with cells. At maturity the cells in the centre of the neck disintegrate to create the channel. The channel is filled with mucilage that results from the breaking down of the cells that initially occupied the channel.

A fertilized egg in an archegonium develops into the sporophyte. The sporophyte consists of a spore-containing capsule which, depending on the species, may be stalked or stalkless. Each spore contains a mix of genes from the two parents and on successful germination will give rise to a new gametophyte.

Bryophyte antheridia are fairly uniform in structure and the same is true for the archegonia. The antheridia vary in size and shape (from globose to somewhat cylindrical) depending on species, but the diagram above captures the essence of any antheridium - a short, narrow stalk supporting a swollen, sperm-producing organ. Similarly, the archegonia vary in size, and relative lengths of the neck, the venter and the length of the supporting foot - but the diagram above shows the essential features of all archegonia.

Individual antheridia and archegonia are microscopic but at times you can see where they are formed. In this photo of the moss *Rosulabryum billardieri* each yellow ball is a cluster of antheridia. The same is the case in this photo of a thallose liverwort in the genus Fossombronia.

Though there is much uniformity at the structural level, there is variety in the formation and arrangement of the archegonia and antheridia.

Once an antheridium has matured and contains viable sperm, the sperm need to get to the eggs in archegonia. The first step for the sperm to get out of the antheridia and the second is to then travel to the archegonia and fertilize the eggs within. Water is essential for both steps.

In some bryophytes a mature antheridium will hold free sperm, but more commonly that's not the case. Rather, each sperm is still held within the cell in which it formed. In such a case, when an antheridium opens, those sperm-containing cells are released and it is only at some stage after release from the antheridium that the single sperm, within each such cell, is liberated. Such liberation may take place shortly after the opening of the antheridium or as long as 15 minutes later, depending on species. In the following, the expression "sperm mass" will mean either a mass of free sperm or a mass of sperm-containing cells, when it's not essential to distinguish the two.

When a mature antheridium is moistened the cells at the apex absorb water, swell and finally burst or open in some way. The sperm mass inside a mature antheridium is under pressure. So, once an antheridium has opened, the sperm mass is forced out. In some bryophytes the force is enough to shoot the sperm mass into the air, allowing dispersal over a relatively wide area. However, in most cases the sperm mass simply oozes into the area around the antheridium and further dispersal is by some other means. While the entire sperm mass may sometimes be released during the forceful extrusion, release is more often a two-stage process. Typically a large percentage of the spore mass is quickly forced out by the built-up internal pressure, but a proportion remains within the antheridium and exits more slowly, over many minutes.

The summary in the previous paragraph is enough to give you a quick grasp of the broad features of the process, but there is variation in the finer detail between species. The liberation and dispersal of sperm matter looks more closely at some of the steps in a few bryophytes. The sperm-to-egg process has been thoroughly studied in a relatively small number of bryophytes. Thus, the examples given on that page may not explain the processes in all bryophytes, but you will at least see some of the variations that are known to occur.

Once an egg has been fertilized the development of the sporophyte begins. The fertilized egg elongates and after a few cell divisions begins to differentiate. The lower portion usually becomes a foot that penetrates the gametophyte and anchors the embryonic sporophyte to the gametophyte. The upper will develop into the spore-bearing capsule (and also the supporting stalk or seta, in species in which the mature capsule is stalked).

The sporophytes are at least partially dependent on the gametophyte for nutrients. Transfer cells develop at the sporophyte-gametophyte boundary in the majority of bryophytes, but not all. These cells are specialized cells that allow efficient transfer of nutrients from the gametophyte to the sporophyte. In the bryophytes where they do occur they may be formed on the gametophyte, the sporophyte or both. So, combined with the possibility of no transfer cells, there are four possibilities. All hornworts have transfer cells and they form only on the gametophyte. Three of the four possibilities occur in mosses. In the majority of moss genera the transfer cells are found on both the gametophyte and the sporophyte, though they are absent in Sphagnum and in a small number of moss genera they are found only on the sporophyte. A common example of the last is the genus Polytrichum and its close relatives. In the liverworts all four possibilities occur. The leafy liverworts have transfer cells only on the sporophytes. In the complex thallose liverworts the transfer cells are found on both sporophyte and gametophyte. In the simple thalloid liverworts there are examples of all four possibilities.

The gametophyte-sporophyte junction often has a convoluted, maze-like form. This gives a larger surface area (and hence more transfer cells) than would a simple, smooth boundary and so increases the rate at which nutrients can flow to the sporophyte.

After fertilization, the archegonium becomes modified into a protective sheath around the young sporophyte. There are significant differences, in both structure and development, between hornwort, liverwort and moss sporophytes. You can find out more about the external appearance by going to the bryophyte groups section. In the sporophyte development section you'll find more detailed accounts of sporophyte development and their internal structure.

In mosses the archegonia are typically formed in groups. In many cases once one archegonium in such a group has been fertilized the others lose the ability to be fertilized. This appears to be caused by an inhibitory hormone released from a fertilized archegonium. In such circumstances only one sporophyte can develop from that archegonial group. However, in some circumstances more than one sporophyte may develop from an archegonial group. Such a phenomenon is called polysety. It may be due to two archegonia being fertilized simultaneously or perhaps because of too low an amount of inhibitory hormone being produced.

VEGETATIVE REPRODUCTION

Vegetative reproduction involves no mixing of genes from two parents. There is but one parent for the new individual. Many home gardeners are familiar with vegetative reproduction and practise it without necessarily knowing it by that name. A very common example is the growing of a new plant from a cutting. Similarly, many gardeners are familiar with the dividing of plants. A gardener who divides a plant to get two where there was previously one is again propagating vegetatively.

Bryophytes can reproduce vegetatively in various ways. In fact, it would almost be true to say that vegetative reproduction is the rule and not the exception. Vegetative reproduction is known from bryophytes where sexual reproduction has never been seen.

If the older parts of a branching or forking gametophyte die, the younger parts are left as separate individuals. The thalli of liverworts in the genus Riccia typically fork to produce Y-shaped growth forms. In due course the tips of each Y will fork in turn and you can see that beginning to happen in this photo. In Riccia the older parts die and decay, and you can see the early stages in this photo. In particular you can still see a hint of the Y shape on the left, but the thallus at the branching point is brown and dying. Soon you would have two, separate, strap-like thalli that would, in time, form new forks at their younger ends and so form new Y-shaped growth forms. This forking and dying process can lead to a rosette-like growth form for a Riccia colony, as shown in this photo.

A similar process occurs in a number of mosses. Many mosses grow in compact colonies or mingled with other vegetation. In such circumstances, as the stems lengthen and branch the older parts may die and decay, creating separate plants where once there were branched plants.

Most forms of vegetative reproduction involve some part of the gametophyte breaking off, being carried away by some means (e.g. wind, water, animal), landing somewhere else and then developing into a new plant. If a gametophyte fragment lands in a suitable habitat, it will grow into a new gametophyte.

The bits that are capable of growing into new plants vary in form. They may simply be pieces that have been accidentally broken off. They may be ordinary parts of the plant that are pre-disposed to breakage. They may be tiny, specialized bud-like outgrowths that bear no resemblance to any "ordinary" part of the plant. A variety of terms are used to describe the bits that can grow into new plants. For the most part they won't be used here. The important thing is to learn something about the ways in which bryophytes reproduce vegetatively, rather than simply bits of jargon. One widely used expression is vegetative propagule, which is a useful, general term for any piece (large or small) that can produce a new plant vegetatively.

In the form of vegetative reproduction illustrated by Riccia and Sphagnum, the new plants form alongside the original plants. This method produces a fairly slow spread. As noted earlier, some vegetative propagules can be very small and the smaller the fragment the more easily it is carried further away from the parent plant (especially by wind or water).

In many cases of re-growth from a vegetative propagule, a new thallose or leafy plant does not simply grow immediately from the propagule. Rather, a protonema first grows out of the propagule and a new thallose or leafy plant develops from the protonema. A protonema is a juvenile, undifferentiated growth form (for example, the first stage in gametophyte development after spore germination). It often resembles a featureless, green wash on its substrate. In such cases of vegetative reproduction there is a reversion to a juvenile stage before the development of a new thallose or leafy growth

There are many ways in which part of a bryophyte could get broken off. A wombat starting to dig its burrow scrapes a liverwort colony apart and sends pieces off in various directions. In moist forests you will find curtains of mosses, such as the one pictured here. Often these will be high up on the forest trees and may be easily broken by storms or animals. That may lead to strands of moss landing elsewhere. In each of the two cases noted here the dispersed pieces will continue growing if their new locations provide suitable conditions. On a more mundane level think of this moss, growing in a lawn that has been recently mown. You can see the cut ends of the grass blades. The lawn mower may also have cut the moss and flung some fragments further afield.

The scenarios in the previous paragraph involve random accidental or deliberate breakages by some other agency, with no real involvement by the bryophytes. It is important to also note that many bryophytes are quite brittle when dry, so making breakage much easier than in the moist state. However many bryophytes (in particular among the mosses and leafy liverworts) naturally develop weakened areas in parts of the gametophyte. Such areas are pre-disposed to breakage and the consequent liberation of particular (rather than random) gametophyte fragments as vegetative propagules.

Campylopus Clavatus, a Moss

In *Campylopus clavatus*, a tufted moss, a small apical section of each stem is weakly joined to the rest of the stem. This apical section will break off easily, for example if an animal walks across the moss colony or raindrops hit the colony with sufficient force. You will typically see a colony of *Campylopus clavatus* with numerous loose apical fragments lying on top of the colony . Each fragment consists of a very short stem section, with attached leaves, and on the right is a closer view. Such fragments are easily dispersed by wind or water. In fact if you run your finger across such a moss colony you will see those fragments "jump" about, with some landing a little away from the colony. This will happen even if the moss cushion is moist, rather than dry. The moss *Pohlia nutans* has catkin-like branches which are readily detached from the parent plant. Some hornworts in the genus Megaceros have frilly thallus margins which can fragment readily. In the leafy liverwort genus *Pycnolejeunea* even a light touch is enough to break off a shoot fragment with several attached leaves. Plants of the Northern Hemisphere leafy liverwort *Lejeunea cardotii* are prostrate and typically grow as mats on tree trunks or dead wood. Small-leaved branches grow from the large-leaved, main stems of the plant. Those small-leaved branches may be produced in immense numbers, one behind each leaf of a main stem. Furthermore, these small-leaved branches may in turn produce more small-leaved branches. In many colonies of this liverwort the small-leaved branches greatly outnumber the larger-leaved main stems. The small-leaved stems are very fragile and are easily broken off and dispersed as vegetative propagules.

The term deciduous is used to describe such readily-detachable features. Many other leafy liverworts also have deciduous branches and both mosses and leafy liverworts have species with deciduous leaves. In some cases, rather than a whole leaf being deciduous, there's a line of weakness within the leaf so that only a part of the leaf breaks off and disperses readily as a propagule. The leaves in many leafy liverwort species are lobed or with pronounced finger-like projections. In a number of species such lobes or projections break off readily and are dispersed as vegetative propagules.

The northern hemisphere leafy liverwort *Gymnocolea inflata* has deciduous perianths. A perianth is a protective tube, composed of fused leaves, around an egg-bearing archegonium in a leafy liverwort. Archegonia are part of the sexual reproduction cycle. In Gymnocolea inflata the perianths around fertilized archegonia do not become deciduous. That's understandable, for otherwise they would not function to protect the developing sporophyte. However, the perianths around un-fertilized archegonia are easily shed. Moreover, the perianths around unfertilized archegonia become more inflated than the ones around fertilized archegonia. So, when shed, each such inflated perianth contains an air bubble which makes the perianth easily carried by water. The perianths of Chonecolea doellingeri, another Northern Hemisphere leafy liverwort, are also very easily dislodged. In this species the deciduous perianths are not of a different form.

So far the deciduous fragments have been simply parts of the plant in their expected places. It is also possible for what look like ordinary plant parts to grow in unexpected areas. The moss *Pilotrichella flexilis* is found in the northern hemisphere. One report from Mexico describes miniature branches growing from the leaves on normal branches and notes that those miniature branches are easily shed. Normally you'd expect branches to grow out from stems or other branches - not from leaf surfaces. The miniature branches (with their own miniature leaves) were essentially of the same structure as normal branches, just much smaller. Any branchlet found growing from a leaf arose from a single leaf cell and a single leaf could give rise to several branchlets. Such branchlets were also found growing from the ordinary branches of the moss. When the branchlets developed on leaves, they did so only from the cells in the basal corner regions, close to where the ordinary leaves joined the ordinary branches. The cells in that area are technically called alar cells and in many mosses are different from other cells in a leaf. There's more about alar cells and their function in the leaf section. In *Pilotrichella* the branchlets developed directly from the alar cells, without any intermediate protonemal stage. A number of mosses produce miniature branches as propagules, but the *Pilotrichella* case was the first time such a development had been observed from moss leaves. The distance between alar cells and the nearby ordinary branch is quite small. But the bryologist who reported this discovery noted that cells which had become differentiated into leaf cells had to de-differentiate and then genes that initiate branch development had to be triggered. The growth of similar miniature shoots (with miniature leaves) has long been known from leaf cells of liverworts in the genus Plagiochila.

All the methods of vegetative reproduction described so far have involved the shedding of some part of a well-developed feature. We've seen, for example, that new plants can grow from deciduous branch tips, branches, leaves and miniature stems. Many bryophytes produce gemmae. These are vegetative propagules which are not simply standard parts of the sorts of features just mentioned. Rather they are small, largely undifferentiated growths. Gemmae may be formed as outgrowths from some part of the gametophyte, in which case they are called exogenous gemmae. By contrast, endogenous gemmae are formed within gametophyte cells. Exogenous gemmae are attached to the gametophyte during their development but this attachment weakens as the gemmae mature, allowing the mature gemmae to be broken off fairly easily. Endogenous gemmae are released by the breakdown of the surrounding cell. Exogenous gemmae are more common than endogenous gemmae.

You'll see small, circular cups on the gametophytes of thallose liverworts in the genus Marchantia . Within these cups you can see some small green discs. Each of those is an exogenous gemma and may get splashed out by a raindrop or washed out by flowing water. The circular cup is called a gemma cup. The great majority of the bryophytes that produce exogenous gemmae don't do so in specialised structures. Instead gemmae are mostly produced simply as outgrowths from some part of the gametophyte – thalli, stems, leaves, rhizoids – depending on species.

Exogenous gemmae have been found on the rhizoids of over 100 moss species. The fact that rhizoidal gemmae are buried and may function as survival organs (akin to the tubers in various flowering plants) has led to these gemmae also being called tubers. No liverwort or hornwort produces rhizoidal gemmae. Liverworts and hornworts form structures that are also called tubers, but these are not produced on rhizoids. Typically these form as swellings at apices of shoots or lobes. Often the shoot or lobe apex turns towards the earth and grows into the soil, so that the tuberous growth is formed below ground level. Often such a tuberous growth is more a means of survival, rather than a means of reproduction since there is no multiplication of individuals. As with exogenous gemmae, endogenous gemmae may be found in various parts of the gametophyte, depending on the species involved.

New plants may grow from the leaf tips of the moss Octoblepharum albidum. In this development rhizoids and, a little later, buds grow at the leaf tips. The buds produce shoots and leaves which, in turn, may produce additional leaf-tip shoots. Such a phenomenon is known in several fern species, where frond tips that come into contact with the ground may root and develop new fern plants. Such ferns are commonly called "walking ferns". The researchers who reported the phenomenon in Octoblepharum noted that in the field such "walking mosses" constituted from 5 to 20 percent of the plants they saw. In contrast, in the mosses they grew in artificial culture over 9 months, the walking variety accounted for about 50% of the plants. In another moss, Leucobryum glaucum, the terminal rosette of leaves may also bear a tuft of rhizoids. New plants may develop on those rhizoids, fall off and then continue growing where they land. Vegetative propagules which look like plant buds grow from cells in at least two species of the thallose liverwort genus Fossombronia, one from Australia and one from South Africa.

New plants can grow from a great variety of fragments. In laboratory experiments new plants have been produced from more gametophyte parts, such as archegonial necks and stalks, paraphyses, calyptrae, rhizoid fragments and even the scales on the undersides of thallose liverworts (right) in the family Marchantiaceae. Such results suggest that almost any part of a gametophyte can be induced to produce a new plant.

Thus far the propagules that have been mentioned have been from the gametophyte. In laboratory experiments new plants have been produced from fragments of moss setae. Many mosses have stalked spore capsules and such a stalk is technically called a seta. The seta is part of the sporophyte and is diploid, whereas the gametophyte is haploid. In the haploid stage a bryophyte has one set of chromosomes in each cell nucleus, and in the diploid stage two sets per nucleus. One set of chromosomes comes from each parent and are combined in the process of sexual reproduction. A setal fragment will give rise to a new gametophyte, but one with two sets of chromosomes in each cell nucleus. Such a gametophyte is therefore genetically distinct from the original gametophyte from which the seta grew.

It should not be surprising to find that many bryophytes may reproduce vegetatively in more than one way. For example, there's nothing special about random fragmentation, so there's nothing to stop a gemma-producing bryophyte also shedding such fragments. Bryum argenteum has deciduous shoot tips and also produces rhizoidal gemmae. The moss Octoblepharum albidum, can reproduce vegetatively via gemmae as well as via shoots at leaf tips, as described above. The northern-hemisphere thallose liverwort Blasia pusillus produces two different types of gemmae by quite different mechanisms. This appears to be unique in the liverworts. Stellate gemmae are produced directly on the thallus surface while ellipsoidal gemmae are produced within long-necked flasks that develop on the thallus surface. Blasia forms a symbiotic association with Nostoc, a Cyanobacterial genus. Each liberated stellate gemma contains Nostoc and, over the summer months, the stellate gemmae germinate to produce fresh, Nostoc-allied thalli. The stellate gemmae do not survive winter. The ellipsoid gemmae do not contain Nostoc and so, when they germinate after liberation, need to form fresh symbiotic associations with free-living Nostoc cells. The ellipsoid gemmae are liberated mainly in late summer to autumn and are able to survive winter. Amongst the mosses there are species which may produce gemmae on the protonema, rhizoids and on the leafy-stemmed plant. The production of one type of gemma neither entails nor precludes the production of another type.

Once spores or vegetative propagules have been produced they need to be released and dispersed if new plants are to develop. There is considerable variation in sporophyte anatomy – in both the spore capsule and, when present, the supporting seta. All aspects of sporophyte structure have some influence on how the spores get out and are dispersed. The aim of this section is to show you many of the ways in which dispersal can happen and, for spore dispersal, the roles played by sporophyte anatomy. We'll look first at the ways in which spores are dispersed and then at vegetative propagules.

Most bryophytes rely on wind for spore dispersal. The vast majority of species have small spores, typically with diameters of 5 to 50 micrometres, a micrometre being a thousandth of a millimetre. An example at the other extreme is the moss genus Archidium, with spore diameters mostly in the range 100-200 micrometres, but as low as 50 micrometres, in Archidium dinteri (known only from southern Africa) and up to 300 micrometres, almost a third of a millimetre, in Archidium ohioense. The latter is a widespread species, known from Africa, Asia, North America, the West Indies and New Caledonia.

Small spores can be carried considerable distances by the wind. Even very light breezes, virtually imperceptible to a person, can easily waft the smaller spores away. Wind dispersal gets more difficult with spores of about 50 micrometre diameter so that Archidium spores, for example, are too heavy for wind to be an effective dispersal agent. Strong winds may certainly move them short distances, just as sand grains can be blown about, but they would be carried more easily by water. In addition, such spores may well be dispersed when mixed up with mud that is picked up by animal feet. Finally, there is a small number of moss species in which insects are the main agents of spore dispersal.

Regardless of how the spores are dispersed they must first get out of the capsule. The capsule may develop a well-defined mouth, through which the spores can escape. If the capsule lacks such a mouth it may split along well-defined lines of weakness (the dehiscence lines) or break open irregularly to expose the spores, for further dispersal by wind or some other agency.

In mosses the majority of species have capsules with well-defined mouths but you will also find species where the capsules break irregularly and the capsules in a couple of genera have dehiscence lines. The majority of liverwort species have capsules with dehiscence lines but there are also species with disintegrating capsules. Hornwort capsules have one or two dehiscence lines. There is variation in the structure of mouths and the ways of splitting.

In a few moss genera the capsule disintegrates and examples of this are Acaulon, Archidium, Ephemerum and Pleuridium. The spores either tumble out of the broken capsules or may be washed away, for example by flowing surface water after rain. The genus *Pleurophascum* (confined to the southern coast of Western Australia, Tasmania and the south island of New Zealand) also appears to have disintegrating capsules, though there are still some unanswered questions about this genus. The identity of the creature responsible for the grazing is unknown as is the role, if any, that this creature plays in spore dispersal. The capsules of the endemic New Zealand species *Pleurophascum ovalifolium* are globose when immature but (unlike those of Pleurophascum grandiglobum) collapse to a discoid shape when mature. The capsules of this species seem to take much longer to develop than do those of the Tasmanian species and they also appear to be longer lasting. One New Zealand bryologist has speculated that the entire spore capsule, when close to maturity but still globose, may function as a dispersal agent. The globose capsule contains much air and could easily float on water and would presumably disintegrate, and release spores, at some distance from the parent plant. In this connection it is worth noting *Pleurophascum ovalifolium* characteristically occurs in very wet sites.

In the complex thallose liverwort genus Riccia the spore capsules are embedded in the thallus. Riccia is a widespread and commonly seen genus, with many species. When mature the capsule and overlying thallus disintegrate, leaving the spores exposed within a cup-like depression. The spores in this genus are commonly 60-80 micrometres in diameter and too large to be easily wind-dispersed, but water could wash them away. Moreover, as the thallus keeps growing at its tip, the older parts will progressively disintegrate. So eventually any spores that have been unable to disperse from those cup-like depressions will be left loose on the soil, where they may germinate or disperse more easily.

Fossombronia, a simple thallose liverwort genus, is also widespread with many species. A mature spore capsule is raised on a flimsy, translucent seta and the capsule wall breaks irregularly into small plate lets, which fall away to expose the spore mass.

At first sight it might appear that complex thallose liverwort genus Targionia has spore capsules that split. This photo shows several plants with mature spore capsules. At the ends of the green, strap-like thalli you can see what look like open, black clam-shells. You could be excused thinking that these are black capsules that have opened to release the spores. In fact those black "shells" are not part of the capsule, though they do surround the developing capsule and form a protective pouch. The capsule itself has thinner walls that break. Targionia is commonly found on soil in habitats that periodically become very dry. Spores may at times escape as the pouch decays. However, there is another, more common process. As conditions dry the thallus closes, the sides rolling inwards, towards the long central axis. Instead of being a green strap, a thallus now looks like a black cord. The black scales that were originally on the underside of the thallus show well after the inrolling. At the same time that black "cord" arches up from the ground to raise the pouch, which opens to expose the spores and elaters from the already ruptured capsule . Thus, even though the spore capsule develops close to the soil, a drying atmosphere raises the pouch (and hence the spores) a centimetre or two into the air where they have a greater chance of being caught and dispersed by breezes.

If we take the point where the capsule is attached to a seta (or, in the absence of a seta, to the gametophyte) as the "south pole" and the opposite point as the "north pole", then the dehiscence lines are oriented north-south like lines of longitude. The number of dehiscence lines varies between species. When a capsule splits along dehiscence lines there are two possibilities – the splitting goes all the way from the "south pole" to the "north pole" or it stops short. In the first case a mature capsule opens out in a number of arms to give a somewhat star-like appearance. This is what occurs in the majority of liverwort species. Usually there are four dehiscence lines and hence four arms in the open capsule. Within the capsules there are elaters as well as spores. Elaters are tubular cells with spiral thickenings that often help in spore release. Elaters do not work in the same way in all species. The elaters may twist or untwist with changes in humidity, or spring suddenly when released from tension. In such cases the movement of the elaters helps fling the spores a short distance into the air where air currents can pick them up and carry them away. In some liverworts the elaters in the spore capsules move about little, if at all, and play little, if any, role in spore release.

Dehiscing capsules may split in the way just described. The other possibility, noted earlier, is that the splitting stops short of the "north pole". In such a case the capsule cannot open out fully, since the arms are joined at their apices. In two closely-related moss genera, Andreaea and Andreaeaobryum, the mature capsule has four or more lines of weakness. In many species of these genera the lines of weakness do not extend to the apex of the capsule. As the mature capsule begins to dry out the capsule shrinks in length. The outer capsule cells shrink less than the inner ones and this causes the capsule to bow out so that slit-like gaps form along the dehiscence lines and the spores can fall out through those gaps. If the capsule is moistened the gaps close up, but will re-open when dry again. Note that a dehiscing liverwort capsule, once open, stays open and does not close up if moistened. When the capsules of the mosses mentioned here are dry and showing the gaps, they look a bit like old-style lanterns - so giving these mosses the common name of Lantern Mosses.

Hornwort spore capsules are generally of a long, tapering form, the exception being the genus Notothylas in which the capsules are relatively short. At maturity hornwort capsules split, along their length, along either one or two dehiscence lines. The splitting starts near, but not at, the apex of the capsule. Depending on whether the capsule has one or two lines of weakness, it opens via one or two slits. The spores near the apex mature first, then the ones a little lower down, then the ones further down and so on. As the spores lower down mature, so the slit (or slits) extend downward, keeping pace with the maturing spores. The capsule becomes twisted as it dries and the slits open to allow spores to be blown out by breezes. Under moist conditions the capsule untwists and the slits close up to block spore release.

In the great majority of mosses the mature spore capsules have well-defined mouths through which the spores are released, The mouths are formed at the end of the spore capsule opposite the point at which the capsule is attached to the seta or, if there is no seta, opposite the point at which the capsule is attached to the gametophyte. During the development of the spore capsule the mouth is covered by a firmly attached lid (or operculum). That attachment must be broken if the spores are to get out.

In Sphagnum the process is typically explosive, with spores and operculum shot off simultaneously. As the mature capsule begins to dry it shrinks, compressing the air inside. Eventually the internal pressure becomes enough to force the operculum off and shoot the spores into the air where breezes will pick them up. In most mosses the process is not explosive. Rather, the operculum is released fairly gently and the spores are released over an extended period. Even in Sphagnum spore release is not always explosive. This genus is most often found in bogs. Sometimes a rise in water levels may leave mature capsules submerged and then the explosive process cannot take place, since it relies on the drying out of the capsule. In such circumstances the capsule falls off its supporting stalk and the columella decays to leave a small hole at the base of the capsule. Spores can escape through that hole. Another possibility is for the spores to germinate while still in the attached capsule and then burst the capsule as the germinating plants expand.

Eccremidium is a predominantly Australian moss genus. They are soil mosses with gametophytes no more than a few millimetres tall and the spores are fairly large, from 50 to 140 micrometres in diameter. The capsules are spherical to pear-shaped with the operculum occupying about half the capsule. This is unusual, with the opercula in other genera occupying very little of the capsule. Once the spores of an Eccremidium have matured the operculum falls off, leaving a smooth-rimmed mouth that is relatively large, often with a diameter equal to that of the spore capsule. In three of the six Eccremidium species known from Australia the seta holding the capsule is bent over so that the capsule is held with the mouth angled downwards. The large spores would find it easy to fall out of the large, smooth-rimmed mouth. Even in species where the mouth is not angled downwards some disturbance of the capsule (for example by wind, water or animal) would probably be enough to shake the spores out.

Here are some plants of the genus Bryum , each with a green, immature capsule atop a seta. The capsules are also held so that the mouths face downward and they will keep this orientation as the capsules mature and turn from green to brown. In each capsule the operculum is relatively small but things still seem simple enough. Once the operculum has come off surely the spores will fall out. However, a closer look shows that things aren't quite that simple. In the majority of mosses (including the genus Bryum) the mouth is lined with teeth of some sort. These are called the peristome teeth by some writers (with the rim around the mouth being the peristome), while others simply use the word peristome to mean a toothed mouth. Peristome teeth may move in response to changes in humidity, either closing or opening the mouth to stop or allow spore release. There is variation in structure of peristome teeth and there are genera which lack peristome teeth. You've already seen Eccremidium as an example of the latter and Sphagnum is another. Given the explosive nature of spore release in Sphagnum, it is clear that such teeth would have no function - and would in fact hinder spore release.

Any raindrop (or runoff from overhead plants) that hits the upper side of the capsule momentarily depresses the capsule wall and so (analogous to a puffball fungus) forces a puff of spores out between those threadlike teeth. Capsules in the genera Buxbaumia and Diphyscium also present relatively large surface areas, though the capsules are smaller than those of Dawsonia, often no more than half a centimetre in length. Once again capsules struck by falling raindrops puff out spores. In the case of Buxbaumia the capsules orient themselves so that the mouth is pointed towards the highest light intensity. Where the light intensity is highest, the obstructions are least. Puffing the spores in that direction would increase their chances of clearing surrounding obstacles and dispersing further away. The painting at the top of this page shows views of Buxbaumia aphylla. On the upper right you can see a close-up of a capsule, in reality about five millimetres long. On the lower left is a much closer view of the peristome and on the right are some whole plants. Between the capsule and peristome pictures is the calyptra, which covers the very young sporophyte.

In a small number of moss species (in the family Splachnaceae) spore dispersal is primarily by dung- or carrion-loving insects. These mosses grow on the dung of various animals and occasionally on old animal carcases. Theoretically the spores are small enough to be wind-dispersed but they are sticky and clump together, so ruling out wind dispersal. The capsules are often highly modified, coloured to attract insects and producing insect-attracting chemicals. Here is a description of the spore release process in some of these mosses. At maturity the spore capsule sheds the operculum. In dry conditions the capsule walls shrink, forcing the peristome teeth to bend back so as to finish up turned down against the outside wall of the spore capsule. At the same time the shrinkage of the capsule leads to the columella extending beyond the capsule mouth. The tip of the columella is coated with the sticky spores. Insects, attracted to the capsule, will almost inevitably pick up clumps of the sticky spores. Being dung- or carrion-loving insects they'll naturally visit other carcases or droppings and so carry spores exactly to the sorts of substrates that these mosses exploit. In moist conditions the capsule swells (so bringing the columella back within the capsule) and the peristome teeth fold back over the mouth and spore release stops.

Schistostega pennata, a widespread Northern Hemisphere moss, is another species with sticky spores. It's not in the family Splachnaceae and also seems to be without any features (such as colour or chemicals) that would attract a specific type of organism to act as a dispersal agent.

There are many agents which can help in the dispersal of vegetative propagules. To take the example closest to home, think of humans. In Such fragments could then be easily carried further afield by that lawn mower. Alternatively, suppose that a gardener is raking fallen leaves off that lawn. The rake may well catch and pull out some strands of this creeping moss - which fall elsewhere as the gathered leaves are being removed. Both the mown fragments and the raked fragments are capable of generating new plants in the right habitats. Thinking of taking a walk through a grassy paddock? One bryologist found fragments of the moss Thuidiopsis furfurosa had adhered to his socks when he'd walked through a grassy, New Zealand meadow. This moss is brittle in the dry state, so fragments could easily break off and attach to fur, feathers - or socks. Most people are well aware of the annoying burrs, grass seeds and so on that are readily picked up by socks. In some grassy areas various species of creeping mosses may grow fairly luxuriantly and, with the surrounding grasses for support, grow to ankle height where they can get caught by socks. On a bush walk you will have brushed against some shrubs or had a lie down. Later that day, as you're about to get in your car for the trip home, you brush bits of rubbish from your jumper – leaves, seeds, twigs and fragments of moss or liverwort. How far have you carried those fragments – 10 metres or 10 kilometres? You've just acted as a very effective disperser of vegetative propagules part from humans many other animals, in their normal activities, may help disperse bryophyte fragments. A German study, published in 2001, found 106 bryophyte fragments on 9 wild boar and 25 roe deer. Those fragments represented 12 species. The bristly coats of wild boar picked up more fragments than the sleeker coats of the roe deer. The wallowing and rooting habits of wild boar make it very easy for them to pick up bryophyte fragments. Deer, when lying down, could pick up fragments on their coats. To study this the researchers used a "dummy deer", made of a deer skin filled with foam plastic. This dummy was placed on its stomach on the forest floor. In addition the researchers mimicked a deer's wallowing motion by gently rocking the dummy from side to side a few times and also by pushing it back and forth with gentle pressure. Then the dummy's skin was cleaned of all adhering plant fragments and those were studied. The whole process was done 300 times, at random points in the forest study site, and the dummy yielded 51 bryophyte fragments. Both the boar and the deer had also picked up fragments in their hooves. Wild boar in particular, with their bristly coats and ranging up to 5 kilometres per day in European forests, may well be significant dispersers of forest bryophytes. This study was a small one, with a very small number of animals examined and there are some interesting unanswered questions. For example, how representative of other deer and boar were these 34 animals? Furthermore, in the course of a day an animal could pick up fragments, drop some of them, pick up some more, drop some more and so on. What is the total number of fragments moved per animal per day? However, as the researchers stated, the subject of animals and bryophyte fragments has not been studied systematically

This study has pointed out some interesting possibilities and shown that further study would be worthwhile. Moreover, think of what could be happening in an Australian setting - a potaroo digging for native truffles, a wombat pushing through undergrowth, two possums fighting on a tree branch, an arid area red kangaroo creating a shallow soil scrape. In each of those situations bryophytes could be fragmented and lodge in animal fur.

Fragments of the cosmopolitan moss species Bryum argenteum have been found on the feet of Antarctic skuas and penguins. Presumably as these birds land on or walk over a mossy patch fragments occasionally get scuffed loose and then get picked up unintentionally. Dense bryophyte cushions create stable micro-habitats for various invertebrates. You can often see insectivorous birds pecking or scraping such cushions to get at those invertebrates. In the process fragments of various sorts may be produced and even picked up accidentally. Various birds deliberately pick up strands of trailing mosses and use them to help camouflage nests.

In Queensland the Spectacled Flying Fox (Pteropus conspicillatus) is potentially occasional disperser of bryophytes. Viable fragments have been collected from the droppings of this bat and grown on in the laboratory in artificial culture. That still leaves open the question of what is the fate of the dung-embedded fragments in the wild, but presumably at least a small proportion would grow into new plants. Rather than deliberately choosing to eat bryophytes the evidence suggests that the bats swallow fragments while grooming.

Numerous invertebrates live in bryophyte colonies or move through them. Various invertebrates eat bryophytes, lay their eggs on them or excavate burrows in them. Some caddis fly larvae use bryophyte fragments on their larval cases. During all such activities small fragments could be accidentally released and of course a bryophyte fragment on a discarded larval case may continue growing if that larval case is discarded in a suitable habitat. The widespread moss species *Fissidens fontanus* (which you may also see referred to as *Octodiceras fontanum*) is found on rocks in and beside streams. In Northern Europe it is also found on dead or live freshwater clams of the species *Anodonta cygnea*. These clams may move occasionally and so help disperse the moss. *Leptodictyon riparium* is another moss that is typically found on streamside rocks but which has also been reported on molluscs. Liverworts or mosses have been found on Papuan weevils and Brazilian harvestmen. It is likely that in the course of their roaming these invertebrates could lose pieces of bryophytes, for example during fights. The bryophytes involved are also found on rocks or plants, so the species are not reliant on the invertebrates

Even the disturbance caused by a small invertebrate moving along a bryophyte colony may be enough to loosen a tiny gemma or a fragile branch tip. The dislodged propagules could simply fall onto the immediate surrounds, but some could be picked up by the passing invertebrate on its furry or bristly body, to be dislodged or groomed off later. The northern hemisphere moss *Schistostega pennata* produces gemmae on the protonemal stage . These gemmae are rounded at the end that is attached to the protonema, but long and tapering at the opposite end. That tapering end is extremely sticky in fresh material and mites have been seen with the gemmae of this moss attached to their legs. Undoubtedly various other invertebrates would also pick up such sticky gemmae. It is interesting to note that the spores of *Schistostega pennata* are also sticky.

Inanimate forces may also break pieces off bryophytes. Storms may break and blow away bryophyte covered twigs. If those twigs land in a suitable habitat the bryophytes can continue growing in their new location. From time to time streamside erosion will break bryophyte colonies, with the stream then carrying any pieces further afield. Once again, if the pieces land in suitable habitats they'll continue growing. Strong winds may cause fragmentation, particularly in areas with little in the way of windbreaks. In desert, alpine and polar regions (where even low shrubbery is sparse to absent) winds may blow unchecked and for long periods. Furthermore, wind-blown sand or snow crystals add to the abrasive effects of wind alone, a sustained wind is drying and dry bryophytes are usually brittle. Putting all these factors together, we have ideal conditions for fragmentation. In many cold regions periods of freezing alternate with periods of thawing and such freeze/ thaw cycles could also cause fragmentation.

In a study of a site on Bathurst Island, in the Canadian Arctic, the researchers estimated that there were at least 4,000 propagules per cubic metre of granular snow near the end of the yearly melt. The particular snow bed being studied had melted completely during the previous summer. Therefore all fragments would have been deposited during the winter immediately before the investigators did their sampling. They also tried growing about 900 fragments back at the laboratory and over a four and a half month period 12% showed new growth. Naturally, there will always be questions as to how accurately a laboratory result represents what happens in nature. However, the study does show that a large number of viable propagules could be produced annually on Bathurst Island. At the other end of the world, windblown vegetative propagules have also been studied from the Antarctic and sub-Antarctic areas. On Macquarie Island or at Casey station in Antarctica researchers found gemmae, deciduous shoots, leaves, leaf fragments and stem fragments with attached leaves. Many of these produced new growth in laboratory experiments.

THE IMPORTANCE OF MOSSES

If the mosses had not survived into the present, we would be forced to invent them as just the sort of intermediate we might expect between essentially aquatic algae and fully terrestrial plants. Mosses and, to a lesser extent, other basal plants, do have differentiated shoots. Although these are generally only a few millimeters tall, they are still designed to provide mechanical support against gravity on land — the first such structure in any kingdom. Bryophytes also have leaves of a sort (phyllids). These are typically one cell thick and lack veins, although they may have a central thickening for support. Mosses also have rhizomes. These may have some function in extracting soil nutrients, although their primary function seems to be mechanical attachment to the substrate. Thus they are not true roots, but do approach that condition.

The bottom line is that, structurally, mosses really differ from rhyniophytes in only one aspect: mosses lack the particular, specialized vascular tissues of tracheophytes. That alone is sufficient to explain the lack of big leaves, long stems, and true roots. This whole complex of characters is thus probably primitive. The other distinctive character of mosses is that the plant we normally observe is the haploid, gametophyte stage. But this character is shared with liverworts (basal embryophytes) and so is also probably plesiomorphic.

Curiously, in hornworts (also basal embryophytes) the sporophyte generation is dominant. In addition, it turns out that the leaves of moss probably evolved independently from the leaves of higher plants. So the relationships of the mosses and basal embryophytes are still uncertain. What really seems to set mosses apart is their unique form of leaf. What may unite mosses with higher plants is (a) the presence of stomata to control water loss and (b) meristem (apical growth) in the sporophyte generation.

A word about terminology. "Bryophyta" was formerly used, and is still used by many to include the liverworts and hornworts, as well as the mosses. Since the branch order and monophyly of the mosses and the two liverworts (Marchantiophyta) and hornworts (Anthocerotophyta) remains unclear, this is understandable. However, we prefer the more restrictive use of the term to mean just mosses, since the mosses are probably monophyletic. The wort + moss group, however structured, is almost certainly paraphyletic. That is, the broader definition takes in all of the basal radiation of embryophytes (land plants), including the direct ancestors of the "higher plants".

As mentioned, the gametophyte (haploid) generation is dominant in the life cycle of mosses. The sporophyte (diploid) form remains attached to and dependent on the gametophyte at all times. That being the case, we can best describe the mosses starting from the bottom and working up to the gametic structures, then proceeding from the archegonium (ovary) to the developing sporophyte as if it were an extension of the gametophyte, which is essentially the case. There is considerable variety among mosses; but, when in doubt, we sometimes defer to the peat moss Sphagnum, which is not only the most basal living moss, but also the most common.

Mosses manage to look like small tracheophytes, but behave more like liverworts. So, for example, mosses have small, threadlike basal processes which look like roots. In fact, these are simply rhizoid holdfasts, not very different from those found in algae. It is likely that the rhizoids absorb water, but only because almost everything in a moss absorbs water. However, a moss lacks a highly specialized vascular system, so the water is usually stored in the ubiquitous moss water storage cells. From these cells, water probably diffuses, very slowly, to other parts of the plant; but, in most cases, the probable function of the water storage cell is structural. Since mosses have no wood, either, they use the turgor pressure of these storage cells to satisfy much of their modest need for structural support. Another common moss strategy for structural support is to grow very densely. So, like academic communities, they grow progressively larger and denser, maintaining their continued existence by sheer bulk without having any real roots in the environment which supports them.

Actually, we may further extend this dubious simile, since mosses have likewise developed elaborate designs for clinging to each other, so that the entire mass behaves like a foam rubber pillow—conforming to the slightest pressure, but springing back to business as usual when the pressure is removed. One such strategy is the use of yet more rhizoids, growing from the shoot, which entangle adjacent shoots and their rhizoids.

Mosses generally have a well-defined shoot in the gametophyte generation. In early development, the plant may take a flattened form, suspiciously similar to the thallus of an alga or liverwort. More typically, the spore develops a long thread-like, recumbent growth, the protoneme, which in turn buds off lateral caulonemata which develop into stem-like gametophore structures on which the gametes will develop in a terminal gametangium. This structure typically has a well-developed pattern of leaves, as in the image of Physcomitrella above. It is common to see this referred to as a "stem" or "shoot" although it has no xylem or phloem and is therefore not really similar to an axis of a vascular plant.

Rather than the xylem or phloem of tracheophytes, mosses have tissues made up of hydroid and leptoid cells. These are frequently referred to as "vascular tissues," and they do probably serve analogous functions. The subject has been reviewed by Ligrone et al. (2000). However, they are not lignified. The hydroid tissues, in particular, are the general run of moss water storage cells and are probably only remotely related to the xylem of tracheophytes. Leptoids are not as well understood, but might be homologous to phloem. However there is great variation in the structure and degree of specialization of these tissues, and their phylogenetic distribution is somewhat unexpected. Unfortunately, we have to break off with these vague and enigmatic generalities, or we will never get on with the rest of the job. As it turns out, most water and nutrient transport is actually accomplished external to the shoot, using a variety of rather ad hoc tools—small channels on the surface, gutters and catchments formed by leaves, playing games with mucous and surface tension, etc.

One more celltype worth mention is the stereid. These are thick-walled celulosic support cells found near the perimeter of the shoot which presumably function as rebr. They are easily seen on the perimeter of the shoot in the cross-section of Neckera in the image above. Growth and differentiation of the shoot, and of nearly everything else in mosses, is driven by a single apical cell.

Leaves and Peripheral Structures

Bryophyte leaves are generally arranged in a very stereotyped way around the shoot, typically a spiral with four positions (e.g. Physcomitrella). The bryophyte leaf typically consists of a single layer of photosynthetic cells with a midline vein, rib, nerve or costa (all the same thing) containing hydroids and/or rows of stereids for mechanical support. Botanists who prefer a more exacting definition of leaf refer to them as phyllids, since moss leaves lack many of the advanced features of tracheophyte leaves. The polytrichid mosses may have much more elaborate structures in both leaves and shoot. By contrast, many mosses lack even the customary costal thickening.

In addition to leaves, moss stems may bear structures for aesexual reproduction—gemmae or bulbils. These are not complex. Mosses have great powers of regeneration and can regenerate an entire plant from an isolated segment of a shoot. The gemmae therefore consist of a small ball of undifferentiated cells. Finally, mosses may, like higher plants, also have a waxy cuticle and stomata to help control evaporation. The stomata are restricted to the sporophyte generation—not only in mosses, but in all land plants, with the exception of the extinct polysporangiophytes—and many mosses lack stomata entirely. Nevertheless, general statements to the effect that mosses lack cuticles and/or stomata are simply incorrect.

Mosses may be either monoecious (hermaphroditic) or dioecious. Similar species within the same genus may differ in this character, and there is little phylogenetic consistency. The gametangia are normally located in terminal clusters at the end of the shoot. However, in Takakia and Sphagnum, generally considered two of the most basal mosses, gametangia are associated with leaves along the shoot. As we might expect in such a regime, the male and female gametangia have somewhat similar structures.

Both gametangia form from a stereotyped series of divisions of a terminal apical cell, unlike the gametangia of liverworts and hornworts. However, once the initial division are complete, the developmental pattern of the archegonium is the same in all three groups. "In all three groups, the process involves three longitudinal divisions that form a central triangular axial cell surrounded by three peripheral cells. The peripheral cells form the neck and venter while the axial cell gives rise to the neck canal cells, ventral canal cell and egg. Further divisions in the peripheral cells typically result in a neck of five or six cell rows".

The following description of the structure and function of the antheridium from Phylum Bryophyta is better than anything we could come up with, so we quote it in extenso: "Antheridia are equally elongated with a long narrow stalk and banana-shaped antheridial body. The sterile jacket is one layered except at the tip where an operculum forms. When mature, the operculum cells dissociate and release blocks of spermatozoids individually enclosed in "vesicles". Spermatozoids are biflagellated [sic], coiled and thread-like [and, we might add, absolutely bizarre in both development and appearance]. Transport of spermatozoids may be effected by rain droplets encountering the cup-like antheridial shoot and splashing packets of cells to adjacent plants." Interestingly, this method of dispersal, by raindrops agitating a cup-like structure, is essentially the same as the method used to disperse gemmae during aesexual reproduction in both moss and other basal plants.

THE SPOROPHYTE

The sporophyte generation of mosses is completely dependent on the gametophyte and never makes independent contact with the substrate. Accordingly, we treat it as an anatomical extension of the gametophyte. A generalized bryophyte sporophyte can be considered as having three parts: a "foot" which connects the embryo to the gametophyte to extract nutrients, a seta which lifts the mature sporophyte clear of the gametophyte body, and a terminal sporangium capsule.

As the sporophyte grows, it elongates and grows wider, eventually fracturing the calyptra. The last remnants form a loose, pointed apical dunce cap, the calyptra, and sometimes a ring around the base, the vaginula. At the same time (approximately — the timing is quite variable) three other things are going on. the inner sporophyte capsule (a/k/a urn or sporangium) is developing and hardening. The sporophyte is levitating itself clear of the mother-ship on a long extension tube, the seta Inside the sporophyte, cells on the surface of a central columella (think of it as a little internal shoot) are undergoing meiotic divisions to create haploid spores. The small spaces where this is going on are called loculi. Loculus may be translated to mean, oddly enough, "small space." Perhaps fortunately, people tend to avoid Latin these days and refer to it as a locule.

Later, the calyptra dries up completely and falls off, and the capsule is exposed, with the end sealed by a small operculum. Eventually, this, too, is lost. Many derived mosses have yet another control on spore dispersal: a peristome with peristome teeth arranged around an annulus. The teeth are quite diagnostic of the species and are much sought after by species-level taxonomists for this reason. The teeth are so constructed that they bend outward as they dry out, causing the opening to widen and release the spores. Note that that this entire system is designed to promote spore dispersal in relatively sustained dry weather and moderate winds. That seems an odd sort of requirement for a plant that requires flowing water for sexual reproduction.

Bear in mind that that this describes sporophyte development in an imaginary "typical" moss. As we keep saying, there is a great deal of variation. Sphagnum, for example, has no seta. In Sphagnum and other basal mosses, the sporophyte is raised by the pseudopodium, a structure produced by the parent gametophyte.

These last three paragraphs will get you by the usual, garden-variety botany mid-term. Because we are kindly and avuncular, but mostly because you must be quite desperate if you are trying to prepare for an examination using this site, we will also add that most teachers of botany have their own, favorite moss—the one they learned bryology on.

Here we have deliberately picked a mixed bag of Latin endings: Sphagnum, "Andreaeids", Polytrichales, and Bryopsida. In a cladistic scheme, it makes no sense to agonize over the rank of the groups. Instead we've simply used a variety of rank suffixes to convey other information — or possibly just to be obnoxious. In any case, Sphagnum is a single genus, so there's no sense in dressing it up as anything else. The Andreaeids are a paraphyletic collection of (probably) three small genera, two of which may have a good deal in common. We've added Takakia because it has no obvious place anywhere else. Polytrichales is a moderately diverse bunch of "typical" mosses—probably monophyletic. Bryopsida is an extremely speciose, but fairly uniform, set of plants—a large and distinct monophyletic assemblage. Consequently, we've given it the highest "rank" consistent with common usage.

Sphagnum is different from all other mosses in many ways. There has never been much doubt that it is the sister group of all other mosses. Sphagnum is the moss which forms peat. There are 250-400 living species which collectively have the distinction of covering 2-3% of the entire land surface of the Earth (an oft-repeated statistic—but how was this computed?). The plants create an environment which is antiseptic and acidic, thus the plants (and anything in them) are extremely slow to decompose.

Many of Sphagnum's unique features may be plesiomorphic ("primitive") traits. The fossil record of mosses is so poor that we can't tell if these are unique specializations or traits left over from the distant common ancestor of all mosses. For example, the phyllids contain a very high proportion of hydroids mixed in with the photosynthetic cells—to the point that the plants often look grey-green rather than the bright green of other common mosses. The spore expulsion (dehiscence) is explosive. These characters, and Sphagnum's acid, antiseptic secretions, appear highly specialized. However, unlike other mosses, Sphagnum does not retain a calyptrum "cap." You can see the naked operculum in the image. Sphagnum sporophytes also lack a peristome. On germination, the spores may (not always) form a structure which looks suspiciously like the flat thallus of a liverwort. As mentioned above, Sphagnum forms a pseudopod from the gametophyte, rather than a seta from the sporophyte.

One deserves special mention because of its possible implications. In all mosses except Sphagnum, the spores themselves are formed from the central, invaginating endothecium tissues—very much in the manner of mesodermal structures developing from the developing coelom in the gastrula embryo of animals. In Sphagnum, the spores develop from the outer amphithecium—just as, in some basal, "acoelomate" animals normally "mesodermal" tissues are derived from ectoderm . . . Compare the Sphagnum sporophyte (the leftmost image in the collection above) with a typical microlecithal animal embryo at gastrula, and ponder that one a while. This is undoubtedly a case of convergence, but moderately mind-blowing all the same.

The Andreaeidae, as generally described, include Andreaea and Andreaeobryum. In most respects they created a smooth and continuous path of character acquisition from Sphagnum up through the Bryopsida. The genus Takakia was originally considered to be a very peculiar liverwort. However, when the sporophyte of Takakia was ultimately described. This dug a raw and unseemly pot-hole in the phylogenetic pathway, since Takakia has an odd mix of characters. The current best guess is that Takakia is closest to the Andreaeidae—but it is perhaps more bryopsid-like in some ways.

As mentioned, the Andreaeidae themselves are more cleanly intermediate forms. The gametophyte is almost indistinguishable from the gametophyte of the Polytrichales. However, the sporophyte has a number of primitive features. As in Sphagnum, the gametophytic coat over the sporophyte, the epigonium, is retained until just before dehiscence. Similarly, they the sporophytes are elevated on a pseudopodium, rather than a seta. Unlike Takakia or bryopsids, Andreaea has a relatively short sporophyte foot.

However, the foot in Andreaeobryum is elongate. The Andreaeidae are also primitive in (probably) producing spores from endothecial tissue, although this tissue grows over the outer wall from the base of the columella, rather than covering the columella as in Sphagnum. The capsule wall itself is even less specialized than in Sphagnum. It rests directly on the spore mass, is single-layered, and lacks a distinct operculum. Dehiscence in all Andreaeids is accomplished by slitting the lateral walls of the capsule.

Our previous tiresome remarks notwithstanding, botanists have drawn a real distinction between Polytrichopsida and Polytrichales in that the former includes the Tetraphidales, encompassing such genera as Buxbaumia, Oedipodium, Tetraphis, and sometimes even Andreaeobryum and Takakia. Hyvönen et al. . The possible inclusion of the last two genera strongly signals the likelihood that this is a garbage taxon. So, for the moment, we will ignore it and collapse Polytrichopsida into Polytrichales.

Polytrichales includes about 19 genera and perhaps 200 species. Polytrichales are typically pioneer plants of open, sometimes even dry, habitats. The group exhibits great diversity: from miniature plants with reduced leaves such as Pogonatum pensilvanicum of eastern North America, to giants of Australasia and New Zealand like *Dawsonia superba* (height up to 50 cm), with the best-developed gametophyte of all land plants. The most typical features of the gametophyte are closely spaced adaxial lamellae on the leaves, forming a pseudoparenchyma, and differentiation of leaves into a distinct blade and sheathing base. The calyptra is typically hairy in many common species of the Northern Hemisphere, enveloping the developing capsules of the sporophyte generation. This has given the whole group its name, although most genera have a practically naked calyptra. Capsules of the Polytrichales normally have a well-developed peristome with at least 16 teeth formed of whole cells.

The Bryopsida include 90% of all moss species. Given this fact, and given the long history of the group, it ought to be a simple matter to list the characteristics which identify these as the definitive moss lineage. Thus, bryopsid sporophytes generally have vertically aligned plates embedded between multiple layers of cells in the amphithecium. They possess stomata which may or may not be homologous with those in higher plants (but Sphagnum also has stomata of a sort). The sporophytes also develop opercula, peristomes, continuous columella, and a spongy layer between the amphithecium and the spore mass. The setae frequently twist at some stage of sporophyte development.

However, only one characteristic seems to be both unique and universal among bryopsids. All bryopsid gametophytes are arthrodontous or descend from ancestors which were arthrodontous. That is, these mosses have peristome teeth which are formed by walls growing between the rows of cells making up the mouth of the spore capsule. This and other critical concepts in moss dentistry are explained and illustrated at the Tree of Life page on Bryopsida, thereby saving us the bother.

Most of the general features of moss evolution are implicit in the descriptions above, but we include two odd speculations that may be worth the effort of explaining.

First, we have, somewhat speciously, compared the sporophyte to the gastrula of an animal embryo. Those similarities probably reflect the inherent advantages, and limited number of ways to create, an additional population of specialized developmental cells which can do their job in a tightly controlled internal environment. The best geometrical solution to that problem is probably the one represented by the gastrula or columellar sporophyte. This creates a controlled volume, closely bordered by two layers of different types of cells. One might evolve such a system simply by having a flat embryo—or sporophyte, as the case may be. However, the advantage of creating the space by localized invagination is that it automatically creates proximo-distal polarity, and thus a fundamental tool for differentiation between separate parts of the organism. Although plants never went this direction, the mechanism also creates the further possibility of dorsoventral polarity, if the invagination can be consistently flattened in a particular plane. In mosses the "mesodermal" compartment was co-opted almost completely for the production and storage of gametes. The same is true of many basal invertebrates. However, the elaborate peristomal structures of mosses show that mosses developed the potential of this system to some extent. Evolution is driven by low-probability random events. There does not need to be a "reason" for matters to arrange themselves in any particular way. Still, we wonder why the potential of the system was not expressed in further specializations.

Second, it is a little bit unusual to see this kind of evolutionary tree—with a few, sometimes strongly divergent basal forms and an enormous radiation of a single derived lineage. This signals an unusually poor fossil record, which is indeed the case with mosses, and/or a very old taxon. Either can result in a "patchy" record of basal forms, with most of the early diversity completely unknown. One of the oddest things about the whole embryophyte story is, in fact, its antiquity. The first known embryophyte remains are, if correctly identified, from the Middle Ordovician, or about 470 Ma. It is usually thought that complex plant communities date from the Earliest Devonian (415 Ma) and forests from perhaps 390 Ma. This is not unreasonably slow. After all, the vertebrates probably subsisted as an only moderately successful type of oddly specialized worm for longer than 60 Ma. However, unlike vertebrates, plants were exploiting an entirely new environment. Typically, this sort of evolutionary invasion leads to explosive radiation. If mosses, liverworts, and perhaps a slew of similar forms, developed early, why didn't things move faster? We, certainly, have no answer. However, we are struck with the fact that, in the Late Ordovician, an ice age was triggered by (it is thought) a sudden draw-down in atmospheric carbon dioxide. This rapid reduction in carbon dioxide has, thus far, eluded explanation. Could these matters be related? A similar event in the Mississippian was almost certainly due to the growth of inland forests . Even the "Snowball" episodes of the Cryogenian may have begun with the spread of eukaryotic algae into fresh water (small volume, but very rich in sunlight, nitrates and iron). After all, no law states that every invasion, military or evolutionary, must necessarily result in a successful or permanent conquest. History, and life, are recorded by the winners.

3

Bryophyta

INTRODUCTION

Diversity of mosses has been classified in approximately 10,000 species, 700 genera, and about 110-120 families. This places the mosses as the third most diverse group of land plants, only after the angiosperms and ferns. Mosses are small plants requiring stereoscopes and compound microscopes for routine examination. The conspicuous green leafy shoots are the gametophytes, haploid organisms, on which the diploid embryo develops into a mature sporophyte. The sporophyte is chlorophyllose and photosynthetic only in early stages of development, and it is mostly dependent on the gametophyte. Moss colonies are a very important element in many ecosystems, from the tundra to the tropical rain forest, reducing soil erosion, capturing water and nutrients, providing shelter for microfauna, and nurseries for seedlings in succession or regeneration processes.

As a lineage, mosses are a historically crucial group in the understanding of th transition to life on land. The green leafy shoots (game tophytes) retain some features of the green algal ances tors (chlorophyll a and b, starch, sperm with two forward undulipodia), but the needle-like shoots that produce the spores (sporophy tes) display key innovations for the life outside water, such as stomates, a simple strand of conductive cells [in an unbranched sporophyte], and airborne spores produced in a single apical capsule (sporangium). This is the simplest structural level among all land plants. The next organizational level is found in two fossil groups: Horneophythopsida and Aglaophyton (Rhynia) major, where the sporophyte is branched and produces several sporangia. The sporophyte shows the most complex structural organization in the tracheophytes.

There is no controversy that the mosses are monophyletic. Synapomorphies for the mosses are: (i) leaves in the gametophyte; these green laminar organs are attached densely along the shoot; (ii) multicellular rhizoids; these branched filaments composed of a series of multiple cells develop from the surface of the gametophyte axis at the point of contact with the substrate; (iii) columnella; this is a cylinder of sterile cells located in the center of the capsule. Other possible synapomorphies are features of male gamete [spermatozoid] ultrastructure

The life cycle of a moss alternates a conspicuous gametophyte generation and a dependent sporophyte generation. Some of the salient features of the different life cycle stages are outlined below:

Protonema. Spores germinate and produce a protonema. This is usually filamentous and branched, but in some groups it is tallose or massive. At several places in the protonema, apical cells differentiate and produce the foliose shoots.

Gametophore. An apical cell produces a stem and leaves spirally arranged. The stems produce branches in several combinations of monopodial and sympodial architectures. Leaves are sessile, unlobed, and often with a thickened midrib. Other appendages to the stem are multicellular rhizoids, axillary hairs, paraphyllia, pseudoparaphyllia, and various types of asexual propagules. The patterns of the leaf cell network and leaf cell papillae provide numerous characters for the systematic arrangement of genera and species.

Gametangia. Archegonia and antheridia are produced in groups, with paraphyses among them, and surrounded by perichaetial or perigonial leaves. Autoicous, synoicous, paroicous, heteroicous, and dioicous sexual conditions are found in many taxa.

Seta. The sporophyte consists of the foot, a seta and an apical sporangium. The foot is embedded within the apex of a stem or branch. Ultrastructural details of the transfer zone provide important similarities with other land plants. The seta is short or elongate. It has an internal conductive system that connects the foot and the capsule at both extremes.

SPORANGIUM

A single sporangium or capsule develops distally from an unbranched sporophyte. The sporangium opens by an apical pore, longitudinal splits or most commonly by an operculum. The external cell layer (exothecium) often has stomata, especially in the neck. Underlaying the exothecium there are parenchymatose cells. Both concentric layers constitute the amphithecium . Internally, the endothecium consists of a cylinder of sporogenous tissue, surrounding a columnella of sterile cells. Moss spores are unicellular, sometimes retaining tetrad marks.

PERISTOME

In the majority of mosses, the apex of the capsule (operculum) falls off at maturity and reveals a structure called the peristome. This is a ring of narrow triangular segments surrounding the mouth of the capsule. Changes in moisture conditions cause movements of the peristome and facilitate the dispersion of spores in favorable dry conditions. Two basic types of peristomes are found in mosses: arthrodontous and nematodontous. In arthrodontous peristomes, at the level of the capsule mouth and above, three innermost rings of cells of the amphithecium are involved in the formation of teeth in most taxa. These three concentric rows are known as the "outer", "primary" and "inner" peristomial layers (OPL, PPL, IPL). Most peristomate mosses have the arthrodontous type, in which each tooth is composed of periclinal (tangential) cell wall remnants between two of the three concentric peristomial cell layers (Figure 3.1). If the teeth are formed by the tangential

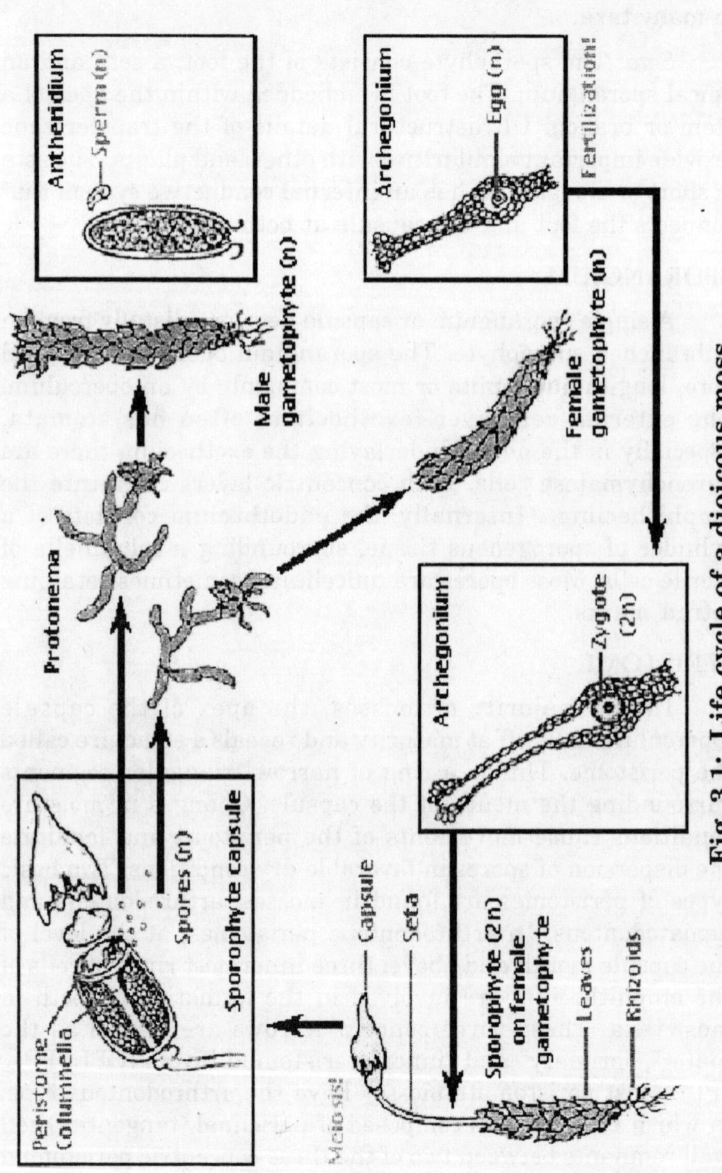

Fig. 3.1: Life cycle of a dioicous moss

walls between the OPL and PPL, the row of teeth is collectively known as the exostome. In the second case, the cell wall remnants are located between the cell rings of the PPL and IPL, therefore the row of segments is known as endostome. A second fundamental type of peristome is nematodontous, structured by narrow columns of entire cell wall remnants. Each tooth consists of agglomerated cylinders formed by the periclinal and anticlinal thickened cell walls. As in arthrodontous peristomes, peristomial cells derive from the innermost amphithecium, but multiple concentric peristomial layers (four to seven) contribute to the formation of a nematodontous tooth.

Early proposals of phylogenetic relationships within the mosses interpreted three fundamental levels of structural organization, largely based on the morphology of the capsules. In the most basal lineage, the Sphagnopsida, capsules open through an apical pore. In the Andreaeopsida, a very small intermediate group, capsules open through longitudinal valves. In the most derived and diverse lineage, the "true mosses" (commonly classified as the Bryopsida), the capsules are operculate and differentiate a peristome. Characters from the exostome and endostome provided a basis for classification of major groups at the ordinal level and above.

Modern explicit hypotheses for moss relationships are based on several cladistic analyses based on morphological, ultrastructural, and molecular data. A recent issue of the journal "The Bryologist" includes a number of new phylogenetic analyses that incorporate exemplars from nearly all families of mosses. Each paper was a major collaborative effort by researchers from many countries who brought together sequence data from rbcL, rps4, trnL-F, and 18S rRNA. Newton et al. presented a cladistic analysis of all orders of mosses based on morphological characters and four DNA sequence data sets for 33 exemplar taxa and ten outgroups. Results include the monophyly of the mosses, the inclusion of Takakia as a moss sister to Sphagnum, and the monophyly of the peristomate mosses. In combination, these recent higher level analyses provide the first cladistic framework for the major lineages of mosses.

Five main lineages are currently recognized, although their taxonomic ranking as classes or subclasses is still controversial. In most cases, higher-level classification of the mosses is not fully settled because there are different names used for the same major clades. Here we use names for these five lineages considered at the rank of class. A basal split divides the Sphagnopsida sister to a large clade in which the Andreaeopsida is the most basal lineage.

Early schemes of relationships within mosses have not changed substantially, but they have been expanded to accommodate a few groups now recognized at higher taxonomic rankings. Ambuchania, a separate genus closely related to Sphagnum, is now classified in its own order and placed as sister to the Sphagnales (Andreaeobryum, a genus related to Andreaea, is also classified in its own order and placed as sister to the Andreaeales).

Ideas about the relationships within the peristomate clade have oscillated between two views. In some systems, the Polytrichopsida were considered as the most advanced because of the complexity of the gametophyte. In contrast, other arrangements interpreted this group as the most ancestral due to the simplicity of the nematodontous peristome. Modern cladistic analyses corroborated the basal position of the nematodontous mosses (Polytrichum) relative to the more recent origin of the diplolepidous mosses (Bryum). However, the phylogenetic positions of certain groups are still controversial. The Tetraphidopsida, for example, has been placed as sister to the Bryopsida or sister to the Polytrichopsida and Bryopsida together. Another example of controversial phylogenetic position is the Oedipodiaceae. This family is alternatively placed as sister to the large clade of peristomate mosses (Polytrichopsida, Tetraphidopsida and Bryopsida) or as sister to just the Polytrichales.

THE FASCINATING BRYOPHYTE

Takakia deserves a special note. Initially it was interpreted as a liverwort, and until recently it was classified in the Calobryales. After the discovery of plants with sporophytes, now it is undoubtedly classified as a moss. This placement within the mosses was confirmed with recent cladistic analyses based on morphology and several genes, although it remains uncertain if it should be sister to Andreaeaopsida or Sphagnopsida.

Several factors have made possible such new insight on higher level relationships within mosses. New empirical data have accumulated on morphology, anatomy, ontogeny, ultrastructure, and DNA sequences. Although most of the recent studies at higher levels are based on molecular data, the most important element of the recent progress has been the use of cladistic methodology for the interpretation of such types of evidence. A formal framework of higher-level moss phylogeny is now available, which replaces former intuitive estimates of relationships founded largely on concepts of overall similarity and ad hoc evolutionary scenarios.

4

Hornworts

INTRODUCTION

Hornworts are a group of bryophytes, or non-vascular plants, comprising the division Anthocerotophyta. The common name refers to the elongated horn-like structure, which is the sporophyte. The flattened, green plant body of a hornwort is the gametophyte plant.

Hornworts may be found world-wide, though they tend to grow only in places that are damp or humid. Some species grow in large numbers as tiny weeds in the soil of gardens and cultivated fields. Large tropical and sub-tropical species of *Dendroceros* may be found growing on the bark of trees.

The plant body of a hornwort is a haploid gametophyte stage. This stage usually grows as a thin rosette or ribbon-like thallus between one and five centimeters in diameter. Each cell of the thallus usually contains just one chloroplast per cell. In most species, this chloroplast is fused with other organelles to form a large pyrenoid that both manufactures and stores food. This particular feature is very unusual in land plants, but is common among algae.

Many hornworts develop internal mucilage-filled cavities when groups of cells break down. These cavities are invaded by photosynthetic cyanobacteria, especially species of *Nostoc*.

Such colonies of bacteria growing inside the thallus give the hornwort a distinctive blue-green color. There may also be small *slime pores* on the underside of the thallus. These pores superficially resemble the stomata of other plants.

The horn-shaped sporophyte grows from an archegonium embedded deep in the gametophyte. Hornworts sporophytes are unusual in that the sporophyte grows from a meristem near its base, instead of from its tip the way other plants do. Unlike liverworts, most hornworts have true stomata on the sporophyte as mosses do. The exceptions are the genera *Notothylas* and *Megaceros*, which do not have stomata.

When the sporophyte is mature, it has a multicellular outer layer, a central rod-like columella running up the center, and a layer of tissue in between that produces spores and pseudo-elaters. The pseudo-elaters are multi-cellular, unlike the elaters of liverworts. They have helical thickenings that change shape in response to drying out, and thereby twist in and thereby help to disperse the spores. Hornwort spores are relatively large for bryophytes, measuring between 30 and 80 µm in diameter or more. The spores are polar, usually with a distinctive Y-shaped tri-radiate ridge on the proximal surface, and with a distal surface ornamented with bumps or spines.

The life of a hornwort starts from a haploid spore. In most species, there is a single cell inside the spore, and a slender extension of this cell called the *germ tube* germinates from the proximal side of the spore. The tip of the germ tube divides to form an octant of cells, and the first rhizoid grows as an extension of the original germ cell. The tip continues to divide new cells, which produces a thalloid protonema. By contrast, species of the family Dendrocerotaceae may begin dividing within the spore, becoming multicellular and even photosynthetic before the spore germinates. In either case, the protonema is a transitory stage in the life of a hornwort.

From the protonema grows the adult gametophyte, which is the persistent and independent stage in the life cycle. This stage usually grows as a thin rosette or ribbon-like thallus between one and five centimeters in diameter, and several layers of cells in thickness. It is green or yellow-green from the chlorophyll in its cells, or bluish-green when colonies of cyanobacteria grow inside the plant.

When the gametophyte has grown to its adult size, it produces the sex organs of the hornwort. Most plants are monoicous, with both sex organs on the same plant, but some plants (even within the same species) are dioicous, with separate male and female gametophytes. The female organs are known as archegonia (singular archegonium) and the male organs are known as antheridia (singular antheridium). Both kinds of organs develop just below the surface of the plant and are only later exposed by disintegration of the overlying cells.

The biflagellate sperm must swim from the antheridia, or else be splashed to the archegonia. When this happens, the sperm and egg cell fuse to form a zygote, the cell from which the sporophyte stage of the life cycle will develop. Unlike all other bryophytes, the first cell division of the zygote is longitudinal. Further divisions produce three basic regions of the sporophyte.

At the bottom of the sporophyte (closest to the interior of the gametophyte), is a foot. This is a globular group of cells that receives nutrients from the parent gametophyte, on which the sporophyte will spend its entire existence. In the middle of the sporophyte (just above the foot), is a meristem that will continue to divide and produce new cells for the third region. This third region is the capsule. Both the central and surface cells of the capsule are sterile, but between them is a layer of cells that will divide to produce pseudo-elaters and spores. These are released from the capsule when it splits lengthwise from the tip.

While the fossil record of crown group hornworts only begins in the upper Cretaceous, the lower Devonian *Horneophyton* may represent a stem group to the clade, as it possesses a sporangium with central columella not attached at the roof. However, the same form of columella is also characteristic of basal moss groups, such as the Sphagnopsida and Andreaeopsida, and has been interpreted as a character common to all early land plants with stomata.

Hornworts were traditionally considered a class within the Division Bryophyta (bryophytes). However, it now appears that this group is paraphyletic, so the hornworts tend to be given their own division, called Anthocerotophyta. The Bryophyta is now restricted to include only mosses.

There is a single class of hornworts, called Anthocerotopsida, or traditionally Anthocerotae. This class includes a single order of hornworts (Anthocerotales) in this classification scheme. In some other classification schemes, a second order Notothyladales (containing only the genus *Notothylas*) is recognized because of the unique and unusual features present in that group.

Among land plants, hornworts appear to be one of the oldest surviving lineages; cladistic analysis implies that the group originated prior to the Devonian, around the same time as the mosses and liverworts. There are only about 100 species known, but new species are still being discovered. The number and names of genera are a current matter of investigation, and several competing classification schemes have been published since 1988.

A study of molecular, ultrastructural and morphological data has yielded a new classification of hornworts given below:

Families and genera

Leiosporocerotaceae

- *Leiosporoceros*

Anthocerotaceae

- *Anthoceros*
- *Folioceros*
- *Sphaerosporoceros*

Notothyladaceae

- *Notothylas*
- *Phaeoceros*
- *Paraphymatoceros*

- *Hattorioceros*
- *Mesoceros*

Phymatocerotaceae

- *Phymatoceros*

Dendrocerotaceae

- *Dendroceros*
- *Megaceros*
- *Nothoceros*
- *Phaeomegaceros.*

Most people are unaware of hornworts, though they are a quite common and widespread group of plants which may be found in tropical forests, along streamsides, and in disturbed fields around the world. Most species are small and unassuming greasy blue-green patches, but some tropical species can cover large areas of soil or the sides of trees. Because of their unique morphology, there have been many questions about relationships and origin of the Anthocerotophyta.

The group's common name "hornwort" refers to the tall narrow sporophytes which are embedded in the top of the plant. As in other bryophytes, the sporophyte remains attached to its parent gametophyte throughout its life, but unlike these other plants, the sporophyte continues to grow throughout its life; this happens as a group of cells at the base of the horn divide repeatedly. This continuous growth from a near-basal meristem is unique among plants to hornworts.

As the sporophyte grows longer, it splits into two halves lengthwise, releasing the spores as they mature. Along with the spores are cells called pseudoelaters, which change shape as they dry out and thereby assist in scattering the spores.

When a spore germinates, it produces a flat thalloid plant with a greasy blue-green colour and odd morphology. The best way to recognize a hornwort, and especially to tell it apart from a liverwort or fern gametophyte, is to look at the plant under a low-power microscope; hornworts will generally have a single large chloroplast per cell. This is an especially useful character for sterile material, but you must look at *fresh* material, since hornworts tend to dry up and fall apart if not kept moist, and will quickly rot if immersed.

The oldest hornwort fossils are spores from the late Cretaceous (Maastrichtian), which have been compared to the spores of the living genus *Phaeoceros*. A somewhat older fossil, *Notothylacites filiformis*, has been described as a hornwort from the Santonian (Late Cretaceous) of Czechoslovakia, but the identification of the fossil as a hornwort is not entirely certain—the fossil preserves the whole plant, but in rather poor detail, and might be better interpreted as a liverwort. Most hornwort fossils are Miocene spores found in Europe and assigned to the Anthocerotaceae.

Such a late date for oldest fossils of hornworts is curious, considering that most cladistic analyses of living plants suggests that the group originated early in the history of land plants, probably before the Devonian, and should be one of the oldest lineages of plants. The presence of a columella, or small column of tissue, running up through the center of the capsule is reminiscent of mosses and of "rhyniophytes", such as *Horneophyton*, and further supports the idea of an early divergence. The delicate hornworts may not have been easily preserved, which might explain the absence of a Paleozoic record. However, it seems puzzling that their spores, which would be much more likely to be preserved, are not encountered either.

CERATOPHYLLUM

Ceratophyllum is a cosmopolitan genus of flowering plants, commonly found in ponds, marshes, and quiet streams in tropical and in temperate regions. They are usually called hornworts, although this name is also used for unrelated plants of the division Anthocerotophyta.

Ceratophyllum grows completely submerged, usually, though not always, floating on the surface, and does not tolerate drought. The plant stems can reach 1–3 m in length. At intervals along nodes of the stem they produce rings of bright green leaves, which are narrow and often much-branched. The forked leaves are brittle and stiff to the touch in some species, softer in others. The plants have no roots at all, but sometimes they develop modified leaves with a rootlike appearance, which anchor the plant to the bottom. The flowers are small and inconspicuous, with the male and female flowers on the same plant. In ponds it forms thick buds in the autumn that sink to the bottom which give the impression that it has been killed by the frost but come spring these will grow back into the long stems slowly filling up the pond.

Hornwort plants float in great numbers just under the surface. They offer excellent protection to fish-spawn, but also to snails, infected with bilharzia. Because of their appearance and their high oxygen production, they are often used in freshwater aquaria.

Ceratophyllum is considered unique enough to warrant its own family, Ceratophyllaceae, and its precise relationship to other angiosperms remains unclear. It was considered a relative of Nymphaeaceae and included in Nymphaeales in the Cronquist system but recent research has shown that it is not closely related to Nymphaeaceae or any other extant plant family. Some early molecular phylogenies suggested it was the sister group to all other angiosperms, but more recent ones have suggested that it is the sister group to either the monocots or the eudicots. The APG II system places the family in its own order, the Ceratophyllales.

The division of the genus into species is not completely settled. More than 30 species have been described, but many are probably just variants of these more widely accepted species:

- *Ceratophyllum demersum* L. (Rigid Hornwort or Common Hornwort)

- *Ceratophyllum echinatum* A.Gray (Spineless Hornwort)
- *Ceratophyllum muricatum* Cham. (Prickly Hornwort)
- *Ceratophyllum platyacanthum* Cham.
- *Ceratophyllum submersum* L. (Soft Hornwort or Tropical Hornwort)

Of these, *Ceratophyllum demersum* is widespread, with a global distribution; the others all have more restricted ranges.

ALTERNATION OF GENERATIONS

The Alternation of phases (or generations) describes the life cycle of plants, fungi and protists. A multicellular diploid phase alternates with a multicellular haploid phase. The term can be confusing for people familiar only with the life cycle of a typical animal. A more understandable name would be "alternation of phases of a single generation" because we usually consider a generation of a species to encompass one complete life cycle. The life cycle of organisms with "alternation of generations" is characterized by each phase consisting of one of two distinct organisms: a gametophyte (thallus (tissue) or plant), which is genetically haploid, and a sporophyte (thallus or plant), which is genetically diploid. A haploid plant of the gametophyte generation produces gametes by mitosis. Two gametes (originating from different organisms of the same species or from the same organism) combine to produce a zygote, which develops into a diploid plant of the sporophyte generation. This sporophyte produces spores by meiosis, which germinate and develop into a gametophyte of the next generation. This cycle, from gametophyte to gametophyte, is the way in which all land plants and many algae.

It is often stated that the distinction of "free-living" is important, because all sexually reproducing organisms can be thought to involve alternating phases, at least at the cellular level as meiosis. However, alternation of generations implies that both the diploid and haploid stages are multicellular and this is more important than "free-living". Such a distinction changes the concept to one separating animals and plants. The gametophyte and sporophyte are usually separate, independent organisms in basal algae such as *Ulva lactuca*, where the gametes are free-swimming, and the zygote is formed in the water. By contrast, in land plants the sporophytes are to a greater or lesser extent dependent on the gametophytes, and *vice versa*, especially in the gymnosperms and angiosperms. Indeed it is a defining characteristic of the land plants, or embryophytes (and hence the name), that a developing multicellular sporophyte is, for at least the first stages of its development, nurtured by the gametophyte, as can be seen most clearly in the bryophytes.

All plants have diploid sporophyte and haploid gametophyte stages that are multicellular, and the differences between plant groups are in the relative sizes, forms, and trophic abilities of the gametophyte or sporophyte forms, as well as the level of differentiation in the gametophytes. An example would be comparing pollen and ovules to bisexual gametophyte thalli. Both approaches are discussed in this article.

Life cycles in which there is no multicellular diploid phase are referred to as haplontic. Life cycles with alternating haploid and diploid phases are diplohaplontic, but the equivalent terms diplobiontic, haplodiplontic, or dibiontic are also in use. Two main types of diplohaplontic (alternating) life-cycles are recognized: if the sporophyte and the gametophyte generations are more or less identical in form, the life cycle is said to be isomorphic, meaning "same form". If the generations have very different morphology, the life cycle is called heteromorphic meaning "different forms".

Heterogamy is a term used to describe alternation between parthenogenic and sexually reproductive phases that occurs in some animals. Although conceptually similar to "alternation of generations", the genetics of heterogamy is significantly different.

Fungal mycelia are typically haploid. When mycelia of different mating types meet, they produce two multinucleate ball-shaped cells, which join via a "mating bridge". Nuclei move from one mycelium into the other, forming a *heterokaryon* (meaning "different nuclei"). This process is called *plasmogamy*. Actual fusion to form diploid nuclei is called *karyogamy*, and may not occur until sporangia are formed. Karogamy produces a diploid zygote, which is a short-lived sporophyte that soon undergoes meiosis to form haploid spores. When the spores germinate, they develop into new mycelia.

Protists

Some protists undergo an alternation of generations, including the slime molds, foraminifera, and many marine algae.

The life cycle of slime molds is very similar to that of fungi. Haploid spores germinate to form swarm cells or *myxamoebae*. These fuse in a process referred to as *plasmogamy* and *karyogamy* to form a diploid zygote. The zygote develops into a plasmodium, and the mature plasmodium produces, depending on the species, one to many fruiting bodies containing haploid spores.

Foraminifera undergo a heteromorphic alternation of generations between a haploid *gamont* and a diploid *agamont* phases. The single-celled haploid organism is typically much larger than the diploid organism.

Alternation of generations occurs in almost all marine algae. In most red algae, many green algae, and a few brown algae, the phases are isomorphic and free-living. Some species of red algae have a complex triphasic alternation of generations. Kelp are an example of a brown alga with a heteromorphic alternation of generations. Species from the genus *Laminaria* have a large sporophytic thallus that produces haploid spores which germinate to produce free-living microscopic male and female gametophytes.

Bryophyte plants including the liverworts, hornworts and mosses undergo an alternation of generations; the gametophyte generation is the most common. The haploid gametophyte produces haploid gametes in multicellular gametangia. Female gametangia are called archegonium (pl. archegonia) and produce eggs, while male structures called antheridium (pl. antheridia) produce sperm. Water is required so that the sperm can swim to the archegonium, where the eggs are fertilized to form the diploid zygote. The zygote develops into a sporophyte that is dependent on the parent gametophyte. Mature sporophytes produce haploid spores by meiosis in sporangia (sing. sporangium). When a spore germinates, it grows into another gametophyte. They are also seedless.

Ferns and their allies, including clubmoss and horsetails, reproduce via an alternation of generations. The conspicuous plant observed in the field is the diploid sporophyte. This plant creates by meiosis single-celled haploid spores which are shed and dispersed by the wind (or in some cases, by floating on water). If conditions are right, a spore will germinate and grow into a rather inconspicuous plant body called a prothallus. The haploid prothallus does not resemble the sporophyte, and as such ferns and their allies have a heteromorphic alternation of generations.

In seed plants, the sporophyte is the dominant multicellular phase, and the gametophytes are both strongly reduced in size and different in morphology. Female gametophytes occur only in the seeds, and male gametophytes only in the pollen. A gymnosperm seed, growing on the diploid sporophyte parent, initially contains a haploid female gametophyte bearing a haploid egg cell enclosed in a cup-shaped structure known as the archegonium. The egg is fertilised by a sperm nucleus from a pollen grain, that contains a miniature male gametophyte. The resulting diploid zygote develops into the seed embryo, which is the diploid sporophyte of the next generation. During its development, which in gymnosperms may take 2-3 years, the offspring sporophyte and its parent gametophyte are both nurtured by the 'grandparent' sporophyte until the seed is ripe enough to be released.

EVOLUTIONARY HISTORY OF PLANTS

Plants have evolved through increasing levels of complexity, from the earliest algal mats, through bryophytes, lycopods, ferns and gymnosperms to the complex angiosperms of today. While the simple plants continue to thrive, especially in the environments in which they evolved, each new grade of organisation has eventually become more "successful" than its predecessors by most measures. Further, most cladistic analyses (although these are often at odds!) suggest that each "more complex" group arose from the most complex group at the time.

Evidence suggests that an algal scum formed on the land 1,200 million years ago, but it was not until the Ordovician period, around 500 million years ago, that land plants appeared. These began to diversify in the late Silurian period, around 420 million years ago, and the fruits of their diversification are displayed in remarkable detail in an early Devonian fossil assemblage known as the Rhynie chert. This chert preserved early plants in cellular detail, petrified in volcanic springs. By the middle of the Devonian period most of the features recognised in plants today are present, including roots, leaves and seeds. By the late Devonian, plants had reached a degree of sophistication that allowed them to form forests of tall trees. Evolutionary innovation continued after the Devonian period. Most plant groups were relatively unscathed by the Permo-Triassic extinction event, although the structures of communities changed. This may have set the scene for the evolution of flowering plants in the Triassic (~200 million years ago), which exploded the Cretaceous and Tertiary. The latest major group of plants to evolve were the grasses, which became important in the mid Tertiary, from around million years ago. The grasses, as well as many other groups, evolved new mechanisms of metabolism to survive the low CO_2 and warm, dry conditions of the tropics over the last 10 million years.

Land plants evolved from chlorophyte algae, perhaps as early as 510 million years ago their closest living relatives are the charophytes, specifically Charales. Assuming that the Charales' habit has changed little since the divergence of lineages, this means that the land plants evolved from a branched, filamentous, haplontic alga, dwelling in shallow, fresh water perhaps at the edge of desiccating pools.

A late Silurian sporangium. Green: A spore tetrad. Blue: A spore bearing a trilete mark – the *Y*-shaped scar. The spores are about 30-35 µm across.

Plants weren't the first photosynthesisers on land, though: consideration of weathering rates suggests that organisms were already living on the land 1,200 million years ago. These organisms were probably small and simple, forming little more than an "algal scum".

The first evidence of plants on land comes from trilete spores,[from the mid-Ordovician (early Llanvirn, ~470 million years ago). The microstructure of the earliest spores resembles that of modern liverwort spores, suggesting they share an equivalent grade of organisation.

Trilete spores are the progeny of spore tetrads. These consist of four identical, connected spores, produced when a single cell undergoes meiosis. Spore tetrads are borne by all land plants, and some algae. Depending exactly when the tetrad splits, each of the four spores may bear a "trilete mark", a *Y*-shape, reflecting the points at which each cell was squashed up against its neighbours. However, in order for this to happen, the spore walls must be sturdy and resistant at an early stage. This resistance is closely associated with having a desiccation-resistant outer wall—a trait only of use when spores have to survive out of water. Indeed, even those embryophytes that have returned to the water lack a resistant wall, thus don't bear trilete marks A close examination of algal spores shows that none have trilete spores, either because their walls are not resistant enough, or in those rare cases where it is, the spores disperse before they are squashed enough to develop the mark, or don't fit into a tetrahedral tetrad.

The earliest megafossils of land plants were thalloid organisms, which dwelt in fluvial wetlands and are found to have covered most of an early Silurian flood plain. They could only survive when the land was waterlogged.

Once plants had reached the land, there were two approaches to desiccation. The bryophytes avoid it or give in to it, restricting their ranges to moist settings, or drying out and putting their metabolism "on hold" until more water arrives. Tracheophytes resist desiccation. They all bear a waterproof outer cuticle layer wherever they are exposed to air (as do some bryophytes), to reduce water loss—but since a total covering would cut them off from CO_2 in the atmosphere, they rapidly evolved stomata—small openings to allow gas exchange. Tracheophytes also developed vascular tissue to aid in the movement of water within the organisms and moved away from a gametophyte dominated life cycle.

The establishment of a land-based flora permitted the accumulation of oxygen in the atmosphere as never before, as the new hoardes of land plants pumped it out as a waste product. When this concentration rose above 13%, it permitted the possibility of wildfire. This is first recorded in the early Silurian fossil record by charcoalified plant fossils. Apart from a controversial gap in the Late Devonian, charcoal is present ever since.

Charcoalification is an important taphonomic mode. Wildfire drives off the volatile compounds, leaving only a shell of pure carbon. This is not a viable food source for herbivores or detritovores, so is prone to preservation; it is also robust, so can withstand pressure and display exquisite, sometimes sub-cellular, detail.

All multicellular plants have a life cycle comprising two phases (often confusingly referred to as "generations"). One is termed the gametophyte, has a single set of chromosomes (denoted 1n), and produces gametes (sperm and eggs). The other is termed the sporophyte, has paired chromosomes (denoted 2n), and produces spores. The two phases may be identical, or phenomenally different.

The overwhelming pattern in plant evolution is for a reduction of the gametophytic phase, and the increase in sporophyte dominance. The algal ancestors to land plants were almost certainly haplobiontic, being haploid for all their life cycles, with a unicellular zygote providing the 2n stage. All land plants (i.e. embryophytes) are diplobiontic – that is, both the haploid and diploid stages are multicellular.

There are two competing theories to explain the appearance of a diplobiontic lifecycle.

The interpolation theory (also known as the antithetic or intercalary theory) holds that the sporophyte phase was a fundamentally new invention, caused by the mitotic division of a freshly germinated zygote, continuing until meiosis produces spores. This theory implies that the first sporophytes would bear a very different morphology to the gametophyte, on which they would have been dependent. This seems to fit well with what we know of the bryophytes, in which a vegetative thalloid gametophyte is parasitised by simple sporophytes, which often comprise no more than a sporangium on a stalk. Increasing complexity of the ancestrally simple sporophyte, including the eventual acquisition of photosynthetic cells, would free it from its dependence on a sporophyte, as we see in some hornworts (*Anthoceros*), and eventually result in the sporophyte developing organs and vascular tissue, and becoming the dominant phase, as in the tracheophytes (vascular plants). This theory may be supported by observations that smaller *Cooksonia* individuals must have been supported by a gametophyte generation. The observed appearance of larger axial sizes, with room for photosynthetic tissue and thus self-sustainability, provides a possible route for the development of a self-sufficient sporophyte phase.

The alternative hypothesis is termed the transformation theory (or homologous theory). This posits that the sporophyte appeared suddenly by a delay in the occurrence of meiosis after the zygote germinated. Since the same genetic material would be employed, the haploid and diploid phases would look the same. This explains the behaviour of some algae, which produce alternating phases of identical sporophytes and gametophytes. Subsequent adaption to the desiccating land environment, which makes sexual reproduction difficult, would result in the simplification of the sexually active gametophyte. and elaboration of the sporophyte phase to better disperse the waterproof spores The tissue of sporophytes and gametophytes preserved in the Rhynie chert is of similar complexity, which is taken to support this hypothesis.

In order to photosynthesise, plants must uptake CO_2 from the atmosphere. However, this comes at a price: while stomata are open to allow CO_2 to enter, water can evaporate. Water is lost much faster than CO_2 is absorbed, so plants need to replace it, and have developed systems to transport water from the moist soil to the site of photosynthesis Early plants sucked water between the walls of their cells, then evolved the ability to control water loss (and CO_2 acquisition) through the use of stomata. Specialised water transport tissues soon evolved in the form of hydroids, tracheids, then secondary xylem, followed by an endodermis and ultimately vessels.

The high CO_2 levels of the Silu-Devonian, when early plants were colonising land, meant that the need for water was relatively low in the early days; as CO_2 was withdrawn from the atmosphere by plants, and stored in coal, so more water was lost in its capture, and more elegant transport mechanisms evolved. As water transport mechanisms, and waterproof cuticles, evolved, plants could survive without being continually covered by a film of water. This transition from poikilohydry to homoiohydry opened up new potential for colonisation Plants were then faced with a balance, between transporting water as efficiently as possible and preventing transporting vessels to implode and cavitate.

During the Silurian, CO_2 was readily available, so little water needed expending to acquire it. By the end of the Carboniferous, when CO_2 levels had lowered to something approaching today's, around 17 times more water was lost per unit of CO_2 uptake However, even in these "easy" early days, water was at a premium, and had to be transported to parts of the plant from the wet soil to avoid desiccation. This early water transport took advantage of the cohesion-tension mechanism inherent in water. Water has a tendency to diffuse to areas that are drier, and this process is exacberated when water can be wicked along a fabric with small spaces. In small passages, such as that between the plant cell walls (or in tracheids), a column of water behaves like rubber – when molecules evaporate from one end, they literally pull the molecules behind them along the channels. Therefore, transpiration alone provided the driving force for water transport in early plants. However, without dedicated transport vessels, the cohesion-tension mechanism cannot transport water more than about 2cm, severely limiting the size of the earliest plants This process demands a steady supply of water from one end, to maintain the chains; to avoid exhausing it, plants developed a waterproof cuticle. Early cuticle may not have had pores but did not cover the entire plant surface, so that gas exchange could continue. However, dehydration at times was inevitable; early plants cope with this by having a lot of water stored between their cell walls, and when it comes to it sticking out the tough times by putting life "on hold" until more water is supplied.

In order to be free from the constraints of small size and constant moisture that the parenchymatic transport system inflicted, plants needed a more efficient water transport system. During the early Silurian, they developed specialized cells, which were lignified (or bore similar chemical compounds) to avoid implosion; this process coincided with cell death, allowing their innards to be emptied and water to be passed through them. These wider, dead, empty cells were a million times more conductive than the inter-cell method, giving the potential for transport over longer distances, and higher CO_2 diffusion rates.

The first macrofossils to bear water-transport tubes *in situ* are the early Devonian pretracheophytes *Aglaophyton* and *Horneophyton*, which have structures very similar to the hydroids of modern mosses. Plants continued to innovate new ways of reducing the resistance to flow within their cells, thereby increasing the efficiency of their water transport. Bands on the walls of tubes, in fact apparent from the early Silurian onwards, are an early improvisation to aid the easy flow of water. Banded tubes, as well as tubes with pitted ornamentation on their walls, were lignified and, when they form single celled conduits, are considered to be tracheids. These, the "next generation" of transport cell design, have a more rigid structure than hydroids, allowing them to cope with higher levels of water pressure Tracheids may have a single evolutionary origin, possibly within the hornworts, uniting all tracheophytes (but they may have evolved more than once).

Water transport requires regulation, and dynamic control is provided by stomata. By adjusting the amount of gas exchange, they can restrict the amount of water lost through transpiration.

An endodermis probably evolved during the Silu-Devonian, but the first fossil evidence for such a structure is Carboniferous. This structure in the roots covers the water transport tissue and regulates ion exchange (and prevents unwanted pathogens etc. from entering the water transport system). The endodermis can also provide an upwards pressure, forcing water out of the roots when transpiration is not enough of a driver.

Once plants had evolved this level of controlled water transport, they were truly homoiohydric, able to extract water from their environment through root-like organs rather than relying on a film of surface moisture, enabling them to grow to much greater size. As a result of their independence from their surroundings, they lost their ability to survive desiccation – a costly trait to retain.

During the Devonian, maximum xylem diameter increased with time, with the minimum diameter remaining pretty constant. By the middle Devonian, the tracheid diameter of some plant lineages had plateaued was consider tracheids allow water to be transported faster, but the overall transport rate depends also on the overall cross-sectional area of the xylem bundle itself. The increase in vascular bundle thickness further seems to correlate with the width of plant axes, and plant height; it is also closely related to the appearance of leaves and increased stomatal density, both of which would increase the demand for water.

While wider tracheids with robust walls make it possible to achieve higher water transport pressures, this increases the problem of cavitation Cavitation occurs when a bubble of air forms within a vessel, breaking the bonds between chains of water molecules and preventing them from pulling more water up with their cohesive tension. A tracheid, once cavitated, cannot have its embolism removed and return to service (except in a few advanced angiosperms which have developed a mechanism of doing so). Therefore it is well worth plants' while to avoid cavitation occurring. For this reason, pits in tracheid walls have very small diameters, to prevent air entering and allowing bubbles to nucleate. Freeze-thaw cycles are a major cause of cavitation. Damage to a tracheid's wall almost inevitably leads to air leaking in and cavitation, hence the importance of many tracheids working in parallel

Cavitation is hard to avoid, but once it has occurred plants have a range of mechanisms to contain the damage. Small pits link adjacent conduits to allow fluid to flow between them, but not air – although ironically these pits, which prevent the spread of embolisms, are also a major cause of them. These pitted surfaces further reduce the flow of water through the xylem by as much as 30% Conifers, by the Jurassic, developed an ingenious improvement, using valve-like structures to isolate cavitated elements. These torus-margo structures have a blob floating in the middle of a donut; when one side depressurises the blob is sucked into the torus and blocks further flow Other plants simply accept cavitation; for instance, oaks grow a ring of wide vessels at the start of each spring, none of which survive the winter frosts. Maples use root pressure each spring to force sap upwards from the roots, squeezing out any air bubbles.

Growing to height also employed another trait of tracheids – the support offered by their lignified walls. Defunct tracheids were retained to form a strong, woody stem, produced in most instances by a secondary xylem. However, in early plants, tracheids were too mechanically vulnerable, and retained a central position, with a layer of tough sclerenchyma on the outer rim of the stems. Even when tracheids do take a structural role, they are supported by sclerenchymatic tissue.

Tracheids end with walls, which impose a great deal of resistance on flow vessel members have perforated end walls, and are arranged in series to operate as if they were one continuous vessel. The function of end walls, which were the default state in the Devonian, was probably to avoid embolisms. An embolism is where an air bubble is created in a tracheid. This may happen as a result of freezing, or by gases dissolving out of solution. Once an embolism is formed, it usually cannot be removed; the affected cell cannot pull water up, and is rendered useless.

The size of tracheids is limited as they comprise a single cell; this limits their length, which in turn limits their maximum useful diameter to 80 μm Conductivity grows with the fourth power of diameter, so increased diameter has huge rewards; vessel elements, consisting of a number of cells, joined at their ends, overcame this limit and allowed larger tubes to form, reaching diameters of up to 500 μm, and lengths of up to 10 m.

Vessels first evolved during the dry, low CO_2 periods of the late Permian, in the horsetails, ferns and Selaginellales independently, and later appeared in the mid Cretaceous in angiosperms and gnetophytes. Vessels allow the same cross-sectional area of wood to transport around a hundred times more water than tracheids! This allowed plants to fill more of their stems with structural fibres, and also opened a new niche to vines, which could transport water without being as thick as the tree they grew on Despite these advantages, tracheid-based wood is a lot lighter, thus cheaper to make, as vessels need to be much more reinforced to avoid cavitation.

Leaves today are, in almost all instances, an adaptation to increase the amount of sunlight that can be captured for photosynthesis. Leaves certainly evolved more than once, and probably originated as spiny outgrowths to protect early plants from herbivory.

The rhyniophytes of the Rhynie chert comprised nothing more than slender, unornamented axes. The early to middle Devonian trimerophytes, therefore, are the first evidence we have of anything that could be considered leafy. This group of vascular plants are recognisable by their masses of terminal sporangia, which adorn the ends of axes which may bifurcate or trifurcate. Some organisms, such as *Psilophyton*, bore enations. These are small, spiny outgrowths of the stem, lacking their own vascular supply.

Around the same time, the zosterophyllophytes were becoming important. This group is recognisable by their kidney-shaped sporangia, which grew on short lateral branches close to the main axes. They sometimes branched in a distinctive H-shape. The majority of this group bore pronounced spines on their axes. However, none of these had a vascular trace, and the first evidence of vascularised enations occurs in the Rhynie genus *Asteroxylon*. The spines of *Asteroxylon* had a primitive vasuclar supply – at the very least, leaf traces could be seen departing from the central protostele towards each individual "leaf". A fossil known as *Baragwanathia* appears in the fossil record slightly earlier, in the late Silurian. In this organism, these leaf traces continue into the leaf to form their mid-vein One theory, the "enation theory", holds that the leaves developed by outgrowths of the protostele connecting with existing enations, but it is also possible that microphylls evolved by a branching axis forming "webbing".

Asteroxylon and *Baragwanathia* are widely regarded as primitive lycopods The lycopods are still extant today, familiar as the quillwort *Isoetes* and the club mosses. Lycopods bear distinctive microphylls – leaves with a single vascular trace. Microphylls could grow to some size – the Lepidodendrales boasted microphylls over a meter in length – but almost all just bear the one vascular bundle. (An exception is the branching *Selaginella*.)

The more familiar leaves, megaphylls, are thought to have separate origins – indeed, they appeared four times independently, in the ferns, horsetails, progymnosperms, and seed plants. They appear to have originated from dichotomising branches, which first overlapped (or "overtopped") one another, and eventually developed "webbing" and evolved into gradually more leaf-like structures So megaphylls, by this "teleome theory", are composed of a group of webbed branches – hence the "leaf gap" left where the leaf's vascular bundle leaves that of the main branch resembles two axes splitting In each of the four groups to evolve megaphylls, their leaves first evolved during the late Devonian to early Carboniferous, diversifying rapidly until the designs settled down in the mid Carboniferous.

The cessation of further diversification can be attributed to developmental constraints but why did it take so long for leaves to evolve in the first place? Plants had been on the land for at least 50 million years before megaphylls became significant. However, small, rare mesophylls are known from the early Devonian genus *Eophyllophyton* – so development could not have been a barrier to their appearance. The best explanation so far incorporates observations that atmospheric CO_2 was declining rapidly during this time – falling by around 90% during the Devonian This corresponded with an increase in stomatal density by 100 times. Stomata allow water to evaporate from leaves, which causes them to curve. It appears that the low stomatal density in the early Devonian meant that evaporation was limited, and leaves would overheat if they grew to any size. The stomatal density could not increase, as the primitive steles and limited root systems would not be able to supply water quickly enough to match the rate of transpiration.

Clearly, leaves are not always beneficial, as illustrated by the frequent occurrence of secondary loss of leaves, famously exemplified by cacti and the "whisk fern" *Psilotum*.

Secondary evolution can also disguise the true evolutionary origin of some leaves. Some genera of ferns display complex leaves which are attached to the pseudostele by an outgrowth of the vascular bundle, leaving no leaf gap. Further, horsetail (*Equisetum*) leaves bear only a single vein, and appear for all the world to be microphyllous; however, in the light of the fossil record and molecular evidence, we conclude that their forbears bore leaves with complex venation, and the current state is a result of secondary simplification.

Deciduous trees deal with another disadvantage to having leaves. The popular belief that plants shed their leaves when the days get too short is misguided; evergreens prospered in the Arctic circle during the most recent greenhouse earth. The generally accepted reason for shedding leaves during winter is to cope with the weather – the force of wind and weight of snow are much more comfortably weathered without leaves to increase surface area. Seasonal leaf loss has evolved independently several times and is exhibited in the ginkgoales, gymnosperms and angiosperms. Leaf loss may also have arisen as a response to pressure from insects; it may have been less costly to lose leaves entirely during the winter or dry season than to continue investing resources in their repair.

The early Devonian landscape was devoid of vegetation taller than waist height. Without the evolution of a robust vascular system, taller heights could not be obtained. There was, however, a constant evolutionary pressure to attain greater height. The most obvious advantage is the harvesting of more sunlight for photosynthesis – by overshadowing competitors – but a further advantage is present in spore distribution, as spores (and, later, seeds) can be blown greater distances if they start higher. This may be demonstrated by *Prototaxites*, thought to be a late Silurian fungus reaching eight metres in height.

In order to attain arborescence, early plants needed to develop woody tissue that would act as both support and water transport. To understand wood, we must know a little of vascular behaviour. The stele of plants undergoing "secondary growth" is surrounded by the vascular cambium, a ring of cells which produces more xylem (on the inside) and phloem (on the outside). Since xylem cells comprise dead, lignified tissue, subsequent rings of xylem are added to those already present, forming wood.

The first plants to develop this secondary growth, and a woody habit, were apparently the ferns, and as early as the middle Devonian one species, *Wattieza*, had already reached heights of 8 m and a tree-like habit.

Other clades did not take long to develop a tree-like stature; the late Devonian *Archaeopteris*, a precursor to gymnosperms which evolved from the trimerophytes, reached 30 m in height. These progymnosperms were the first plants to develop true wood, grown from a bifacial cambium, of which the first appearance is in the mid Devonian *Rellimia*. True wood is only thought to have evolved once, giving rise to the concept of a "lignophyte" clade.

These *Archaeopteris* forests were soon supplemented by lycopods, in the form of lepidodendrales, which topped 50m in height and 2m across at the base. These lycopods rose to dominate late Devonian and Carboniferous coal deposits. Lepidodendrales differ from modern trees in exhibiting determinate growth: after building up a reserve of nutrients at a low height, the plants would "bolt" to a genetically determined height, branch at that level, spread their spores and die They consisted of "cheap" wood to allow their rapid growth, with at least half of their stems comprising a pith-filled cavity. Their wood was also generated by a unifacial vascular cambium – it did not produce new phloem, meaning that the trunks could not grow wider over time.

The horsetail *Calamites* was next on the scene, appearing in the Carboniferous. Unlike the modern horsetail *Equisetum*, *Calamites* had a unifacial vascular cambium, allowing them to develop wood and grow to heights in excess of 10 m. They also branched multiple times.

While the form of early trees was similar to that of today's, the groups containing all modern trees had yet to evolve.

The dominant groups today are the gymnosperms, which include the coniferous trees, and the angiosperms, which contain all fruiting and flowering trees. It was long thought that the angiosperms arose from within the gymnosperms, but recent molecular evidence suggests that their living representatives form two distinct groups. It must be noted that the molecular data has yet to be fully reconciled with morphological data, but it is becoming accepted that the morphological support for paraphyly is not especially strong.

This would lead to the conclusion that both groups arose from within the pteridosperms, probably as early as the Permian.

The angiosperms and their ancestors played a very small role until they diversified during the Cretaceous. They started out as small, damp-loving organisms in the understory, and have been diversifying ever since the mid-Cretaceous, to become the dominant member of non-boreal forests today.

Roots are important to plants for two main reasons: Firstly, they provide anchorage to the substrate; more importantly, they provide a source of water and nutrients from the soil. Roots allowed plants to grow taller and faster.

The onset of roots also had effects on a global scale. By disturbing the soil, and promoting its acidification (by taking up nutrients such as nitrate and phosphate they enabled it to weather more deeply, promoting the draw-down of CO with huge implications for climate These effects may have been so profound they led to a mass extinction.

But how and when did roots evolve in the first place? While there are traces of root-like impressions in fossil soils in the late Silurian body fossils show the earliest plants to be devoid of roots. Many had tendrils which sprawled along or beneath the ground, with upright axes or thalli dotted here and there, and some even had non-photosynthetic subterranean branches which lacked stomata. The distinction between root and specialised branch is developmental; true roots follow a different developmental trajectory to stems. Further, roots differ in their branching pattern, and in possession of a root cap. So while Silu-Devonian plants such as *Rhynia* and *Horneophyton* possessed the physiological equivalent of roots, roots – defined as organs differentiated from stems – did not arrive until later. Unfortunately, roots are rarely preserved in the fossil record, and our understanding of their evolutionary origin is sparse.

Rhizoids—small structures performing the same role as roots, usually a cell in diameter—probably evolved very early, perhaps even before plants colonised the land; they are recognised in the Characeae, an algal sister group to land plants. That said, rhizoids probably evolved more than once; the rhizines of lichens, for example, perform a similar role. Even some animals (*Lamellibrachia*) have root-like structures.

More advanced structures are common in the Rhynie chert, and many other fossils of comparable early Devonian age bear structures that look like, and acted like, roots. The rhyniophytes bore fine rhizoids, and the trimerophytes and herbaceous lycopods of the chert bore root-like structure penetrating a few centimetres into the soil. However, none of these fossils display all the features borne by modern roots Roots and root-like structures became increasingly more common and deeper penetrating during the Devonian period, with lycopod trees forming roots around 20 cm long during the Eifelian and Givetian. These were joined by progymnosperms, which rooted up to about a metre deep, during the ensuing Frasnian stage.

The rhizomorphs of the lycopods provide a slightly approach to rooting. They were equivalent to stems, with organs equivalent to leaves performing the role of rootlets A similar construction is observed in the extant lycopod *Isoetes*, and this appears to be evidence that roots evolved independently at least twice, in the lycophytes and other plants.

A vascular system is indispensable to a rooted plants, as non-photosynthesising roots need a supply of sugars, and a vascular system is required to transport water and nutrients from the roots to the rest of the plant. These plants are little more advanced than their Silurian forbears, without a dedicated root system.

By the mid-to-late Devonian, most groups of plants had independently developed a rooting system of some nature. As roots became larger, they could support larger trees, and the soil was weathered to a greater depth. This deeper weathering had effects not only on the aforementioned drawdown of CO_2, but also opened up new habitats for colonisation by fungi and animals.

Roots today have developed to the physical limits. They pen. The narrowest roots are a mere 40 µm in diameter, and could not physically transport water if they were any narrower. The earliest fossil roots recovered, by contrast, narrowed from 3 mm to under 700 μm in diameter; of course, taphonomy is the ultimate control of what thickness.

The efficiency of many plants' roots is increased via a symbiotic relationship with a fungal partner. The most common are arbuscular mycorrhizae (AM), literally "tree-like fungal roots". These comprise fungi which invade some root cells, filling the cell membrane with their hyphae. They feed on the plant's sugars, but return nutrients generated or extracted from the soil, which the plant would otherwise have no access to.

This symbiosis appears to have evolved early in plant history. AM are found in all plant groups, and 80% of extant vascular plants suggesting an early ancestry; a "plant"-fungus symbiosis may even have been the step that enabled them to colonise the land and indeed AM are abundant in the Rhynie chert; the association occurred even before there were true roots to colonise, and is has even been suggested that roots evolved in order to provide a more comfortable habitat for mycorrhizal fungi.

Evolutionary developmental biology (evo-devo) refers to the study of developmental programs and patterns from an evolutionary perspective. It seeks to understand the various influences shaping the form and nature of life on the planet. Evo-devo arose as a separate branch of science only in the last decade. Most of the synthesis in evo-devo has been in the field of animal evolution, one reason being the presence of elegant model systems like *Drosophila*, *C. elegans*, Zebrafish and *Xenopus*. However, in the past couple of decades, a wealth of information on plant morphology, coupled with modern molecular techniques has helped shed light on the conserved and unique developmental patterns in the plant kingdom also.

The origin of the term "morphology" is generally attributed to Johann Wolfgang von Goethe. He was of the opinion that there is an underlying fundamental organisation *(Bauplan)* in the diversity of flowering plants. In his book titled The

Metamorphosis of Plants, he proposed that the *Bauplan* enabled us to predict the forms of plants that had not yet been discovered Goethe also was the first to make the perceptive suggestion that flowers consist of modified leaves.

In the middle centuries, several basic foundations of our current understanding of plant morphology were laid down. Nehemiah Grew, Marcello Malpighi, Robert Hooke, Antonie van Leeuwenhoek, Wilhelm von Nageli were just some of the people who helped build knowledge on plant morphology at various levels of organisation. It was the taxonomical classification of Carolus Linnaeus in the eighteenth century though, that generated a firm base for the knowledge to stand on and expand. The introduction of the concept of Darwinism in contemporary scientific discourse also had had an effect on the thinking on plant forms and their evolution.

Wilhelm Hofmeister, one of the most brilliant botanists of his times, was the one to diverge away from the idealist way of pursuing botany. Over the course of his life, he brought an interdisciplinary outlook into botanical thinking. He came up with biophysical explanations on phenomena like phototaxis and geotaxis, and also discovered the alternation of generations in the plant life cycle.

The past century witnessed a rapid progress in the study of plant anatomy. The focus shifted from the population level to more reductionist levels. While the first half of the century saw expansion in developmental knowledge at the tissue and the organ level, in the latter half, especially since the 1990s, there has also been a strong impetus on gaining molecular information.

Edward Charles Jeffrey was one of the early evo devo researchers of the 20th century. He performed a comparative analyses of the vasculatures of living and fossil Gymnosperms and came to the conclusion that the storage parenchyma has been derived from tracheids. His research focussed primarily on plant anatomy in the context of phylogeny. This tradition of evolutionary analyses of plant architectures was further advanced by Katherine Esau, best known for her book *The Plant Anatomy*. Her work focussed on the origin and development of various tissues in different plants. Working with Cheadle, she also explained the evolutionary specialization of the phloem tissue with respect to its function.

In the meantime, by the beginning of the latter half of 1900s, *Arabidopsis thaliana* had begun to be used in some developmental studies. The first collection of Arabidopsis thaliana mutants were made around 1945 However it formally became established as a model organism only in 1998.

The recent spurt in information on various plant-related processes has largely been a result of the revolution in molecular biology. Powerful techniques like mutagenesis and complementation were made possible in *Arabidopsis* via generation of T-DNA containing mutant lines, recombinant plasmids, techniques like Transposon Tagging etc. Availability of complete physical and genetic maps RNAi vectors, rapid transformation protocols are some of the technologies that have significantly altered the scope of the field. recently, there has also been a massive increase in the genome and EST sequences of various non-model species, which, coupled with the Bioinformatics tools existing today, generate interesting opportunities in the field of plant evo devo research.

The most important model systems in plant development have been Arabidopsis and Maize. Maize has traditionally been the favourite of plant geneticists, while extensive resources in almost every area of plant physiology and development are available for *Arabidopsis*. Apart from these, Rice, *Antirrhinum*, *Brassica*, Tomato are also being used in a variety of studies. The genomes of *Arabidopsis* and Rice have been completely sequenced, while the others are in process. It must be emphasized here that the information from these "model" organisms form the basis of our developmental knowledge. While *Brassica* has been used primarily because of its convenient location in the phylogenetic tree in the mustard family, *Antirrhinum* is a convenient system for studying leaf architecture. Rice has been traditionally used for studying responses to hormones like abscissic acid and gibberelin as well as responses to stress. However, recently, not just the domesticated rice strain, but also the wild strains have been studied for their underlying genetic architectures.

Some people have objected against extending the results of model organisms to the plant world. One argument is that the effect of gene knockouts in lab conditions wouldn't truly reflect even the same plant's response in the natural world. Also, these supposedly *crucial* genes might not be responsible for the evolutionary origin of that character. For these reasons, a comparative study of plant traits has been proposed as the way to go now.

Since the past few years, researchers have indeed begun looking at non-model, "non-conventional" organisms using modern genetic tools. One example of this is the Floral Genome Project, which envisages to study the evolution of the current patterns in the genetic architecture of the flower through comparative genetic analyses, with a focus on EST sequences. Like the FGP, there are several such ongoing projects that aim to find out conserved and diverse patterns in evolution of the plant shape. Expressed sequence tag (EST) sequences of quite a few non-model plants like Sugarcane, Apple, Lotus, Barley, Cycas, Coffee, to name a few, are available freely online The Cycad Genomics Project, for example, aims to understand the differences in structure and function of genes between gymnosperms and angiosperms through sampling in the order Cycadales. In the process, it intends to make available information for the study of evolution of structures like seeds, cones and evolution of life cycle patterns. Presently the most important sequenced genomes from an evo-devo point of view include those of *A.thaliana* (a flowering plant), Poplar (a woody plant), *Physcomitrella patens* (a bryophyte), Maize (extensive genetic information), and *Chlamydomonas reinhardtii* (a green alga). The impact of such a vast amount of information on understanding common underlying developmental mechanisms can easily be realised.

Apart from EST and genome sequences, several other tools like PCR, Yeast two hybrid system, microarrays, RNA Interference, SAGE, QTL mapping etc. permit the rapid study of plant developmental patterns. Recently, cross-species hybridization has begun to be employed on microarray chips, to study the conservation and divergence in mRNA expression patterns between closely related species. Techniques for analyzing this kind of data have also progressed over the past decade. We now have better models for molecular evolution, more refined analysis algorithms and better computing power as a result of advances in computer sciences.

Evidence suggests that an algal scum formed on the land 1200 million years ago, but it was not until the Ordovician period, around 500 million years ago, that land plants appeared. These begun to diversify in the late Silurian period, around 420 million years ago, and the fruits of their diversification are displayed in remarkable detail in an early Devonian fossil assemblage known as the Rhynie chert. This chert preserved early plants in cellular detail, petrified in volcanic springs. By the middle of the Devonian period most of the features recognised in plants today are present, including roots, leaves and seeds. By the late Devonian, plants had reached a degree of sophistication that allowed them to form forests of tall trees. Evolutionary innovation continued after the Devonian period. Most plant groups were relatively unscathed by the Permo-Triassic extinction event, although the structures of communities changed. This may have set the scene for the evolution of flowering plants in the Triassic (~200 million years ago), which exploded the Cretaceous and Tertiary. The latest major group of plants to evolve were the grasses, which became important in the mid Tertiary, from around 40 million years ago. The grasses, as well as many other groups, evolved new mechanisms of metabolism to survive the low CO_2 and warm, dry conditions of the tropics over the last 10 million years.

The meristematic cells give rise to various organs of the plant, and keep the plant growing. The Shoot Apical Meristem (SAM) gives rise to organs like the leaves and flowers. The cells of the apical meristems - SAM and RAM (Root Apical Meristem)- divide rapidly and are considered to be indeterminate, in that they do not possess any defined end fate. In that sense, the meristematic cells are frequently compared to the stem cells in animals, that have an analogous behavior and function.

Diversity in Meristem Architectures

Is the mechanism of being *indeterminate* conserved in the SAM's of the plant world? The SAM contains a population of stem cells that also produce the lateral meristems while the stem elongates. It turns out that the mechanism of regulation of the stem cell number might indeed be evolutionarily conserved. The *CLAVATA* gene *CLV2* responsible for maintaining the stem cell population in *Arabidopsis* is very closely related to the Maize gene *FASCIATED EAR 2(FEA2)* also involved in the same function. Similarly, in Rice, the *FON1-FON2* system seems to bear a close relationship with the CLV signaling system in *Arabidopsis*. These studies suggest that the regulation of stem cell number, identity and differentiation might be an evolutionarily conserved mechanism in monocots, if not in angiosperms. Rice also contains another genetic system distinct from *FON1-FON2*, that is involved in regulating stem cell number. This example underlines the innovation that goes about in the living world all the time.

Genetics screens have identified genes belonging to the KNOX family in this function. These genes essentially maintain the stem cells in an undifferentiated state. The KNOX family has undergone quite a bit of evolutionary diversification, while keeping the overall mechanism more or less similar. Members of the KNOX family have been found in plants as diverse as Arabidopsis, rice, barley and tomato. KNOX-like genes are also present in some algae, mosses, ferns and gymnosperms. Misexpression of these genes leads to formation of interesting morphological features. For example, among members of *Antirrhinae*, only the species of genus *Antirrhinum* lack a structure called spur in the floral region. A spur is considered an evolutionary innovation because it defines pollinator specificity and attraction. Researchers carried out transposon mutagenesis in *Antirrhinum*, and saw that some insertions led to formation of spurs that were very similar to the other members of *Antirrhinae* indicating that the loss of spur in wild *Antirrhinum* populations could probably be an evolutionary innovation.

The KNOX family has also been implicated in leaf shape evolution. One study looked at the pattern of KNOX gene expression in *A.thaliana*, that has simple leaves and *Cardamine hirsuta*, a plant having complex leaves. In *A.thaliana*, the KNOX genes are completely turned off in leaves, but in *C.hirsuta*, the expression continued, generating complex leaves. Also, it has been proposed that the mechanism of KNOX gene action is conserved across all vascular plants, because there is a tight correlation between KNOX expression and a complex leaf morphology.

Evolution of the Meristem Architecture

The meristem architectures do differ between angiosperms, gymnosperms and pteridophytes. The gymnosperm vegetative meristem lacks organization into distinct tunica and corpus layers. They possess large cells called Central Mother Cells in the meristem. In angiosperms, the outermost layer of cells divides anticlinally to generate the new cells, while in gymnosperms, the plane of division in the meristem differs for different cells. However, the apical cells do contain organelles like large vacuoles and starch grains, like the angiosperm meristematic cells.

Pteridophytes, like fern, on the other hand, do not possess a multicellular apical meristem. They possess a tetrahedral apical cell, which goes on to form the plant body. Any somatic mutation in this cell can lead to hereditary transmission of that mutation. The earliest meristem-like organization is seen in an algal organism from group *Charales* that has a single dividing cell at the tip, much like the pteridophytes, yet more simpler. One can thus see a clear pattern in evolution of the meristematic tissue, from pteridophytes to angiosperms. Pteridophytes, with a single meristematic cell; gymnosperms with a multicellular, but less defined organization and finally, angiosperms, with the highest degree of organization. The genetic innovations that contributed to this evolution are yet not clearly known.

Leaves are the primary photosynthetic organs of a plant. Based on their structure, they are classified into two types—

microphylls, that lack complex venation patterns and megaphylls, that are large and with a complex venation. It has been proposed that these structures arose independently Megaphylls, according to the Telome hypothesis, have evolved from plants that showed a three dimensional branching architecture, through three transformations - planation, which involved formation of a planar architecture, webbing, or formation of the outgrowths between the planar branches and fusion, where these webbed outgrowths fused to form a proper leaf lamina. Studies have revealed that these three steps happened multiple times in the evolution of today's leaves

It has been proposed that the before the evolution of leaves, plants had the photosynthetic apparatus on the stems. Today's megaphyll leaves probably became commonplace some 360mya, about 40my after the simple leafless plants had colonized the land in the early Devonian period. This spread has been linked to the fall in the atmospheric carbon dioxide concentrations in the Late Paleozoic era associated with a rise in density of stomata on leaf surface. This must have allowed for better transpiration rates and gas exchange. Large leaves with less stomata would have gotten heated up in the sun's heat, but an increased stomatal density allowed for a better-cooled leaf, thus making its spread feasible.

Factors Influencing Leaf Architectures

Various physical and physiological forces like light intensity, humidity, temperature, wind speeds etc. are thought to have influenced evolution of leaf shape and size. It is observed that high trees rarely have large leaves, owing to the obstruction they generate for winds. This obstruction can eventually lead to the tearing of leaves, if they are large. Similarly, trees that grow in temperate or taiga regions have pointed leaves, presumably to prevent nucleation of ice onto the leaf surface and reduce water loss due to transpiration. Herbivory, not only by large mammals, but also small insects has been implicated as a driving force in leaf evolution, an example being plants of the genus *Aciphylla*, that are commonly found in New Zealand. The now extinct Moas fed upon these plants, and its seen that the leaves have spines on their bodies, which probably functioned to discourage the moas from feeding on them. Other members of *Aciphylla* that did not co-exist with the moas, do not have these spines.

Genetic Evidences for Leaf Evolution

At the genetic level, developmental studies have shown that repression of the KNOX genes is required for initiation of the leaf primordium. This is brought about by *ARP* genes, which encode transcription factors. Genes of this type have been found in many plants studied till now, and the mechanism i.e. repression of KNOX genes in leaf primordia, seems to be quite conserved. Interestingly, expression of KNOX genes in leaves produces complex leaves. It is speculated that the *ARP* function arose quite early in vascular plant evolution, because members of the primitive group Lycophytes also have a functionally similar gene. Other players that have a conserved role in defining leaf primordia are the phytohormone auxin, gibberelin and cytokinin.

One interesting feature of a plant is its phyllotaxy. The arrangement of leaves on the plant body is such that the plant can maximally harvest light under the given constraints, and hence, one might expect the trait to be genetically robust. However, it may not be so. In maize, a mutation in only one gene called *abphyl (Abnormal Phyllotaxy)* was enough to change the phyllotaxy of the leaves. It implies that sometimes, mutational tweaking of a single locus on the genome is enough to generate diversity. The *abphyl* gene was later on shown to encode a cytokinin response regulator protein.

Once the leaf primordial cells are established from the SAM cells, the new axes for leaf growth are defined, one important (and more studied) among them being the abaxial-adaxial (lower-upper surface) axes. The genes involved in defining this, and the other axes seem to be more or less conserved among higher plants. Proteins of the *HD-ZIPIII* family have been implicated in defining the adaxial identity. These proteins deviate some cells in the leaf primordium from the default abaxial state, and make them adaxial. It is believed that in early plants with leaves, the leaves just had one type of surface - the abaxial one. This is the underside of today's leaves. The definition of the adaxial identity occurred some 200 million years after the abaxial identity was established One can thus imagine the early leaves as an intermediate stage in evolution of today's leaves, having just arisen from spiny stem-like outgrowths of their leafless ancestors, covered with stomata all over, and not optimized as much for light harvesting.

How the infinite variety of plant leaves is generated is a subject of intense research. Some common themes have emerged. One of the most significant is the involvement of KNOX genes in generating compound leaves, as in tomato . But this again is not universal. For example, pea uses a different mechanism for doing the same thing Mutations in genes affecting leaf curvature can also change leaf form, by changing the leaf from flat, to a crinky shape like the shape of cabbage leaves. There also exist different morphogen gradients in a developing leaf which define the leaf's axis. Changes in these morphogen gradients may also affect the leaf form. Another very important class of regulators of leaf development are the microRNAs, whose role in this process has just begun to be documented. The coming years should see a rapid development in comparative studies on leaf development, with many EST sequences involved in the process coming online.

A flower is, arguably, one of the most beautiful products of evolution. Flower-like structures first appear in the fossil records some ~130 mya, in the Cretaceous era.

The flowering plants have long been assumed to have evolved from within the gymnosperms; according to the traditional morphological view, they are closely allied to the gnetales. However, recent molecular evidence is at odds to this hypothesis, and further suggests that gnetales are more closely related to some gymnosperm groups than angiosperms and that gymnosperms form a distinct clade to the angiosperms Molecular clock analysis predicts the divergence of flowering plants (anthophytes) and gymnosperms to ~300 myat.

5

Marchantiophyta

INTRODUCTION

The Marchantiophyta are a division of bryophyte plants commonly referred to as hepatics or liverworts. Like other bryophytes, they have a gametophyte-dominant life cycle, in which cells of the plant carry only a single set of genetic information.

It is estimated that there are 6000 to 8000 species of liverworts, though when Neotropical regions are better studied this number may approach 10,000. Some of the more familiar species grow as a flattened leafless thallus, but most species are leafy with a form very much like a flattened moss. Leafy species can be distinguished from the apparently similar mosses on the basis of a number of features, including their single-celled rhizoids. Leafy liverworts also differ from most (but not all) mosses in that their leaves never have a costa (present in many mosses) and may bear marginal cilia (very rare in mosses). Other differences are not universal for all mosses and liverworts, but the occurrence of leaves arranged in three ranks, the presence of deep lobes or segmented leaves, or a lack of clearly differentiated stem and leaves all point to the plant being a liverwort.

Liverworts are typically small, usually from 2-20 mm wide with individual plants less than 10 cm long, and are therefore often overlooked. However, certain species may cover large patches of ground, rocks, trees or any other reasonably firm substrate on which they occur. They are distributed globally in almost every available habitat, most often in humid locations although there are desert and arctic species as well. Some species can be a nuisance in shady green-houses or a weed in gardens.

Most liverworts are small, usually from 2–20 millimetres (0.08–0.8 in) wide with individual plants less than 10 centimetres (4 in) long, so they are often overlooked. The most familiar liverworts consist of a prostrate, flattened, ribbon-like or branching structure called a thallus (plant body); these liverworts are termed *thallose liverworts*. However, most liverworts produce flattened stems with overlapping scales or leaves in three or more ranks, the middle rank being conspicuously different from the outer ranks; these are called *leafy liverworts* or *scale liverworts*

Liverworts can most reliably be distinguished from the apparently similar mosses by their single-celled rhizoids. Other differences are not universal for all mosses and all liverworts; but the lack of clearly differentiated stem and leaves in thallose species, or in leafy species the presence of deeply lobed or segmented leaves and the presence of leaves arranged in three ranks, all point to the plant being a liverwort. In addition, 90% of liverworts contain oil bodies in at least some of their cells, and these cellular structures are absent from most other bryophytes and from all vascular plants. The overall physical similarity of some mosses and leafy liverworts means that confirmation of the identification of some groups can be performed with certainty only with the aid of microscopy or an experienced bryologist.

Liverworts have a gametophyte-dominant life cycle, with the sporophyte dependent on the gametophyte Cells in a typical liverwort plant each contain only a single set of genetic information, so the plant's cells are haploid for the majority of its life cycle. This contrasts sharply with the pattern exhibited by nearly all animals and by most other plants. In the more familiar seed plants, the haploid generation is represented only by the tiny pollen and the ovule, while the diploid generation is the familiar tree or other plant. Another unusual feature of the liverwort life cycle is that sporophytes (i.e. the diploid body) are very short-lived, withering away not long after releasing spores. Even in other bryophytes, the sporophyte is persistent and disperses spores over an extended period.

The life of a liverwort starts from the germination of a haploid spore to produce a protonema, which is either a mass of thread-like filaments or else a flattened thallus. The protonema is a transitory stage in the life of a liverwort, from which will grow the mature gametophore ("gamete-bearer") plant that produces the sex organs. The male organs are known as antheridia (*singular:* antheridium) and produce the sperm cells. Clusters of antheridia are enclosed by a protective layer of cells called the perigonium (*plural:* perigonia). As in other land plants, the female organs are known as archegonia (*singular:* archegonium) and are protected by the thin surrounding perichaetum (*plural:* perichaeta) Each archegonium has a slender hollow tube, the "neck", down which the sperm swim to reach the egg cell.

Liverwort species may be either dioicous or monoicous. In dioicious liverworts, female and male sex organs are borne on different and separate gametophyte plants. In monoicious liverworts, the two kinds of reproductive structures are borne on different branches of the same plant. In either case, the sperm must swim from the antheridia where they are produced to the archegonium where the eggs are held. The sperm of liverworts is *biflagellate*, i.e. they have two tail-like flagellae that aid in propulsion. Their journey is further assisted either by the splashing of raindrops or the presence of a thin layer of water covering the plants. Without water, the journey from antheridium to archegonium cannot occur.

In the presence of such water, sperm from the antheridia swim to the archegonia and fertilisation occurs, leading to the production of a diploid sporophyte. After fertilisation, the immature sporophyte within the archegonium develops three distinct regions: a foot, which both anchors the sporophyte in place and receives nutrients from its "mother" plant, a spherical or ellipsoidal capsule, inside which the spores will be produced for dispersing to new locations, and a seta (stalk) which lies between the other two regions and connects them. When the sporophyte has developed all three regions, the seta elongates, pushing its way out of the archegonium and rupturing it. While the foot remains anchored within the parent plant, the capsule is forced out by the seta and is extended away from the plant and into the air. Within the capsule, cells divide to produce both elater cells and spore-producing cells. The elaters are spring-like, and will push open the wall of the capsule to scatter themselves when the capsule bursts. The spore-producing cells will undergo meiosis to form haploid spores to disperse, upon which point the life cycle can start again.

Today, liverworts can be found in many ecosystems across the planet except the sea and excessively dry environments, or those exposed to high levels of direct solar radiation As with most groups of living plants, they are most common (both in numbers and species) in moist tropical areas. Liverworts are more commonly found in moderate to deep shade, though desert species may tolerate direct sunlight and periods of total desiccation.

Traditionally, the liverworts were grouped together with other bryophytes (mosses and hornworts) in the Division Bryophyta, within which the liverworts made up the class Hepaticae (also called Marchantiopsida). However, since this grouping makes the Bryophyta paraphyletic, the liverworts are now usually given their own division. The use of the division name Bryophyta *sensu latu* is still found in the literature, but more frequently the Bryophyta now is used in a restricted sense to include only the mosses.

Another reason that liverworts are now classified separately is that they appear to have diverged from all other embryophyte plants near the beginning of their evolution. The strongest line of supporting evidence is that liverworts are the only living group of land plants that do not have stomata on the sporophyte generation. Among the earliest fossils believed to be liverworts are compression fossils of *Pallaviciniites* from the Upper Devonian of New York. These fossils resemble modern species in the Metzgeriales. Another Devonian fossil called *Protosalvinia* also looks like a liverwort, but its relationship to other plants is still uncertain, so it may not belong to the Marchantiophyta. In 2007, the oldest fossils assignable to the liverworts were announced, *Metzgeriothallus sharonae* from the Givetian (Middle Devonian) of New York, USA.

Bryologists classify liverworts in the division Marchantiophyta. This divisional name is based on the name of the most universally recognized liverwort genus *Marchantia*. In addition to this taxon-based name, the liverworts are often called Hepaticophyta. This name is derived from their common Latin name as Latin was the language in which botanists published their descriptions of species. This name has led to some confusion, partly because it appears to be a taxon-based name derived from the genus *Hepatica* which is actually a flowering plant of the buttercup family Ranunculaceae. In addition, the name Hepaticophyta is frequently misspelled in textbooks as Hepatophyta, which only adds to the confusion.

The Marchantiophyta is subdivided into three classes:

- The Jungermanniopsida includes the two orders Metzgeriales (simple thalloids) and Jungermanniales (leafy liverworts).
- The Marchantiopsida includes the three orders Marchantiales (complex-thallus liverworts), and Sphaerocarpales (bottle hepatics), as well as the Blasiales (previously placed among the Metzgeriales). It also includes the problematic genus *Monoclea*, which is sometimes placed in its own order Monocleales.
- A third class, the Haplomitriopsida is newly recognized as a basal sister group to the other liverworts; it comprises the genera *Haplomitrium*, *Treubia*, and *Apotreubia*.

It is estimated that there are 6000 to 8000 species of liverworts, at least 85% of which belong to the leafy group.

Economic Importance

In ancient times, it was believed that liverworts cured diseases of the liver, hence the name. In Old English, the word liverwort literally means *liver plant*. This probably stemmed from the superficial appearance of some thalloid liverworts (which resemble a liver in outline), and led to the common name of the group as *hepatics*, from the Latin word *hepaticus* for "belonging to the liver". An unrelated flowering plant, *Hepatica*, is sometimes also referred to as liverwort because it was once also used in treating diseases of the liver. This archaic relationship of plant form to function was based in the "Doctrine of Signatures"

Liverworts have little direct economic importance today. Their greatest impact is indirect, though the reduction of erosion along streambanks, their collection and retention of water in tropical forests, and the formation of soil crusts in deserts and polar regions. However, a few species are used by humans directly. A few species, such as *Riccia fluitans*, are aquatic thallose liverworts sold for use in aquaria. Their thin, slender branches float on the water's surface and provide habitat for both small invertebrates and the fish that feed on them.

SPORE

In biology, a spore is a reproductive structure that is adapted for dispersal and surviving for extended periods of time in unfavourable conditions. Spores form part of the life cycles of many plants, algae, fungi and some protozoans. A chief difference between spores and seeds as dispersal units is that spores have very little stored food resources compared with seeds.

Spores are usually haploid and unicellular and are produced by meiosis in the sporophyte. Once conditions are favorable, the spore can develop into a new organism using mitotic division, producing a multicellular gametophyte, which eventually goes on to produce gametes.

Two gametes fuse to create a new sporophyte. This cycle is known as alternation of generations, but a better term is "biological life cycle", as there may be more than one phase and so it cannot be a direct alternation. Haploid spores produced by mitosis (known as mitospores) are used by many fungi for asexual reproduction.

Spores are the units of *asexual* reproduction, because a single spore develops into a new organism. By contrast, gametes are the units of *sexual* reproduction, as two gametes need to fuse to create a new organism.

The term *spore* may also refer to the dormant stage of some bacteria or archaea; however these are more correctly known as endospores and are not truly spores in the sense discussed in this article. The term can also be loosely applied to some animal resting stages. Fungi that produce spores are known as *sporogenous*, and those that do not are *asporogenous*.

The term derives from the ancient Greek word (*"spora"*), meaning a seed.

Fungi and fungus-like organisms, spores are often classified by the structure in which meiosis and spore production occurs. Since fungi are often classified according to their spore-producing structures, these spores are often characteristic of a particular taxon of the fungi.

In contrast, in many seed plants and heterosporous ferns, only a single product of meiosis will become a megaspore (macrospore), with the rest degenerating.

- Sporangiospores: spores produced by a sporangium in many fungi such as zygomycetes.
- Zygospores: spores produced by a zygosporangium, characteristic of zygomycetes.
- Ascospores: spores produced by an ascus, characteristic of ascomycetes.
- Basidiospores: spores produced by a basidium, characteristic of basidiomycetes.

- Aeciospores: spores produced by a aecium in some fungi such as rusts or smuts.
- Urediospores: spores produced by a uredinium in some fungi such as rusts or smuts.
- Teliospores: spores produced by a telium in some fungi such as rusts or smuts.
- Oospores: spores produced by a oogonium, characteristic of oomycetes.
- Carpospores: spores produced by a carposporophyte, characteristic of red algae.
- Tetraspores: spores produced by a tetrasporophyte, characteristic of red algae.
- Chlamydospores: thick-walled resting spores of fungi produced to survive unfavorable conditions.

By origin during life cycle

- Meiospores: spores produced by meiosis; they are thus haploid, and give rise to a haploid daughter cell(s) or a haploid individual. Examples are the precursor cells of gametophytes of seed plants found in flowers (angiosperms) or cones (gymnosperms).
- Microspores: meiospores that give rise to a male gametophyte, (pollen in seed plants).
- Megaspores (or macrospores): meiospores that give rise to a female gametophyte, (an ovule in seed plants).
- Mitospores (or cònidia, conidiospores): spores produced by mitosis; they are characteristic of Ascomycetes. Fungi in which only mitospores are found are called "mitosporic fungi" or "anamorphic fungi", and are previously classified under the taxon Deuteromycota (See Teleomorph, anamorph and holomorph).

Spores can be differentiated by whether they can move or not.

- *Zoospores:* mobile spores that move by means of one or more flagella, and can be found in some algae and fungi.

- *Aplanospores:* immobile spores that may nevertheless potentially grow flagella.
- *Autospores:* immobile spores that cannot develop flagella.
- *Ballistospores:* spores that are actively discharged from the body of the fungal fruiting body. Most basidiospores are also ballistospores, and another notable example is spores of *Pilobolus*.
- *Statismospores:* spores that are not actively discharged from the fungal fruiting body. Examples are puffballs.

Trilete spores, formed by the dissociation of a spore tetrad, are taken as the earliest evidence of life on land, dating to the mid-Ordovician (early Llanvirn, ~470 million years ago).

Parlance

In common parlance, the difference between a "spore" and a "gamete" (both together called gonites) is that a spore will germinate and develop into a sporeling, while a gamete needs to combine with another gamete before developing further. However, the terms are somewhat interchangeable when referring to gametes.

A chief difference between spores and seeds as dispersal units is that spores have little food storage compared with seeds, and thus require more favorable conditions in order to successfully germinate. (This is not without its exceptions, however: many orchid seeds, although multicellular, are microscopic and lack endosperm, and spores of some fungi in the Glomeromycota commonly exceed 300μm in diameter.) Seeds, therefore, are more resistant to harsh conditions and require less energy to start mitosis. Spores are produced in large numbers to increase the chance of a spore surviving in a number of notable examples.

The endospores of certain bacteria are often simply called "spores," as seen in the 2001 anthrax attacks, where the media called anthrax endospores "anthrax spores." Unlike eukaryotic spores, endospores are primarily a survival mechanism, not a reproductive method, and a bacterium only produces a single endospore.

Diaspores

In the case of spore-shedding vascular plants such as ferns, wind distribution of very light spores provides great capacity for dispersal. Also, spores are less subject to animal predation than seeds because they contain almost no food reserve; however they are more subject to fungal and bacterial predation. Their chief advantage is that, of all forms of progeny, spores require the least energy and materials to produce.

Vascular plant spores are always haploid and vascular plants are either *homosporous* or *heterosporous*. Plants that are *homosporous* produce spores of the same size and type. *Heterosporous* plants, such as spikemosses, quillworts, and some aquatic ferns produce spores of two different sizes: the larger spore in effect functioning as a "female" spore and the smaller functioning as a "male".

Under high magnification, spores can be categorized as either monolete spores or trilete spores. In monolete spores, there is a single line on the spore indicating the axis on which the mother spore was split into four along a vertical axis. In trilete spores, all four spores share a common origin and are in contact with each other, so when they separate, each spore shows three lines radiating from a center pole.

ENDOSPORE

An endospore is a dormant, tough, and non-reproductive structure produced by bacteria from the Firmicute phylum. The primary function of most endospores is to ensure the survival of a bacterium through periods of environmental stress. They are therefore resistant to ultraviolet and gamma radiation, desiccation, lysozyme, temperature, starvation, and chemical disinfectants. Endospores are commonly found in soil and water, where they may survive for long periods of time. Some bacteria produce exospores or cysts instead.

Structure

In contrast to eukaryotic spores, which are produced by many eukaryotes for reproductive purposes, bacteria will produce a single endospore internally. The spore is often surrounded by a thin covering known as the *exosporium*, which overlies the *spore coat*. The spore coat is impermeable to many toxic molecules and may also contain enzymes that are involved in germination. The *cortex* lies beneath the spore coat and consists of peptidoglycan. The *core wall* lies beneath the cortex and surrounds the protoplast or *core* of the endospore. The core has normal cell structures, such as DNA and ribosomes, but is metabolically inactive.

Up to 15% of the dry weight of the endospore consists of calcium dipicolinate within the core, which is thought to stabilize the DNA. Dipicolinic acid could be responsible for the heat resistance of the spore, and calcium may aid in resistance to heat and oxidizing agents. However, mutants resistant to heat but lacking dipicolinic acid have been isolated, suggesting other mechanisms contributing to heat resistance are at work

The position of the endospore differs among bacterial species and is useful in identification. The main types within the cell are terminal, subterminal and centrally placed endospores. Terminal endospores are seen at the poles of cells, whereas central endospores are more or less in the middle. Subterminal endospores are those between these two extremes, usually seen far enough towards the poles but close enough to the center so as not to be considered either terminal or central. Lateral endospores are seen occasionally.

Examples of bacteria having terminal endospores include *Clostridium tetani*, the pathogen which causes the disease tetanus. Bacteria having a centrally placed endospore include *Bacillus cereus*, and those having a subterminal endospore include *Bacillus subtilis*. Sometimes the endospore can be so large the cell can be distended around the endospore, this is typical of *Clostridium tetani*.

Visualising endospores under the light microscope can be difficult due to the impermeability of the endospore wall to dyes and stains. While the rest of a bacterial cell may stain, the endospore is left colourless. To combat this, a special stain technique called a Moeller stain is used. That allows the endospore to show up as red, while the rest of the cell stains blue. Another staining technique for endospores is the Schaffer-Fulton stain, which stains endospores green and bacterial bodies red.

Formation and Destruction

When a bacterium detects environmental conditions are becoming unfavourable it may start the process of sporulation, which takes about eight hours. The DNA is replicated and a membrane wall known as a *spore septum* begins to form between it and the rest of the cell. The plasma membrane of the cell surrounds this wall and pinches off to leave a double membrane around the DNA, and the developing structure is now known as a forespore. Calcium dipicolinate is incorporated into the forespore during this time. Next the peptidoglycan cortex forms between the two layers and the bacterium adds a spore coat to the outside of the forespore. Sporulation is now complete, and the mature endospore will be released when the surrounding vegetative cell is degraded.

Endospores are resistant to most agents which would normally kill the vegetative cells they formed from. Household cleaning products generally have no effect, nor do most alcohols, quaternary ammonium compounds or detergents. Alkylating agents however, such as ethylene oxide, are effective against endospores.

Whilst resistant to extreme heat and radiation, endospores can be destroyed by burning or autoclaving. Exposure to extreme heat for a long enough period will generally have some effect, though many endospores can survive hours of boiling or cooking. Prolonged exposure to high energy radiation, such as xrays and gamma rays, will also kill most endospores.

Reactivation

Reactivation of the endospore occurs when conditions are more favourable and involves *activation*, *germination*, and *outgrowth*. Even if an endospore is located in plentiful nutrients, it may fail to germinate unless activation has taken place. This may be triggered by heating the endospore. Germination involves the dormant endospore starting metabolic activity and thus breaking hibernation. It is commonly characterised by rupture or absorption of the spore coat, swelling of the endospore, an increase in metabolic activity, and loss of resistance to environmental stress. Outgrowth follows germination and involves the core of the endospore manufacturing new chemical components and exiting the old spore coat to develop into a fully functional vegetative bacterial cell, which can divide to produce more cells.

As a simplified model for cellular differentiation, the molecular details of endospore formation have been extensively studied, specifically in the model organism *Bacillus subtilis*. These studies have contributed much to our understanding of the regulation of gene expression, transcription factors, and the sigma factor subunits of RNA polymerase.

Endospores of the bacterium *Bacillus anthracis* were used in the 2001 anthrax attacks. The powder found in contaminated postal letters was composed of extracellular anthrax endospores. Inhalation, ingestion or skin contamination of these endospores, which were technically incorrectly labelled as "spores", led to a number of deaths.

Geobacillus stearothermophilus endospores are used as biological indicators when autoclave is used in sterilization procedures.

MOSSES AND LIVERWORTS

Mosses and liverworts are traditionally classified together in the Division Bryophyta on the basis of their sharing:

- a similar life cycle (alternation of generations).
- similar reproductive organs (antheridia and archegonia).
- lack of vascular tissue (xylem and phloem).

Some 23,000 species of living mosses and liverworts have been identified. These are small, fairly simple, plants usually found in moist locations.

- Liverworts have a thin, leathery body that grows flat on moist soil or, in some cases, the surface of still water.
- Mosses have an erect shoot bearing tiny leaflike structures arranged in spirals.

Neither mosses nor liverworts have any woody tissue so they never grow very large. They have neither xylem nor phloem for the transport of water and food through the plant.

The leafy shoot of mosses is haploid and thus part of the gametophyte generation.

In the common haircap moss, *Polytrichum commune*, there are three kinds of shoots:

- female, which develop archegonia at their tip;
- A single egg forms in each archegonium.
- male, which develop antheridia at their tip;
- Multiple swimming sperm form in each antheridium.
- sterile, which do not form sex organs.

In early spring, raindrops splash sperm from male to female plants. These swim down the canal in the archegonium to the chamber containing the egg. The resulting zygote begins the sporophyte generation.

The Sporophyte Generation

Mitosis of the zygote produces an embryo that grows into the mature sporophyte generation. It consists of:

- a foot, which absorbs water, minerals, and probably some food from the parent gametophyte.
- a stalk, at the tip of which is formed a sporangium.

The sporangium is:

- filled with spore mother cells;
- sealed by an operculum; and
- covered with a calyptra. The calyptra develops from the wall of the old archegonium and so is actually a part of the gametophyte generation. It is responsible for the common name ("haircap moss") of this species.

During the summer, each spore mother cell undergoes meiosis, producing four haploid spores - the start of the new gametophyte generation. Late in the summer, the calyptra and operculum become detached from the sporangium allowing the spores to be released.

These tiny spores are dispersed so effectively by the wind that many mosses are worldwide in their distribution.

If a spore reaches a suitable habitat, it germinates to form a filament of cells called a protonema. Soon buds appear and develop into the mature leafy shoots. So,

- The gametophyte generation is responsible for sexual reproduction
- The sporophyte generation is responsible for dispersal.

Evolutionary Position of the Bryophytes

Evidence from the chloroplast genome

The chloroplasts of mosses and liverworts, like those of all photosynthetic eukaryotes, contain multiple copies of a small genome: circular DNA molecules encoding some - but not all - of the genes needed for:

- their own replication and
- photosynthesis.

The mitochondrial DNA of plants suggests a different evolutionary scenario. The mtDNA of all plants

- including mosses and lycopsids but
- Not liverworts (nor green algae)

Contain certain shared introns. This suggests that:

- mosses and liverworts are not close relatives;
- liverworts and lycopsids are not close relatives;
- liverworts may have been the first group of plants to evolve from algal ancestors.

The endosymbiotic theory concerns the origins of mitochondria and plastids (e.g. chloroplasts), which are organelles of eukaryotic cells. According to this theory, these organelles originated as separate prokaryotic organisms which were taken inside the cell as endosymbionts. Mitochondria developed from proteobacteria (in particular, Rickettsiales or close relatives) and chloroplasts from cyanobacteria.

The endosymbiotic theory was first articulated by the Russian botanist Konstantin Mereschkowski in 1905 Mereschkowsky was familiar with work by botanist Andreas Schimper, who had observed in 1883 that the division of chloroplasts in green plants closely resembled that of free-living cyanobacteria, and who had himself tentatively proposed that green plants had arisen from a symbiotic union of two organisms. Ivan Wallin extended the idea of an endosymbiotic origin to mitochondria in the 1920s. These theories were initially dismissed or ignored. More detailed electron microscopic comparisons between cyanobacteria and chloroplasts (for example studies by Hans Ris, combined with the discovery that plastids and mitochondria contain their own DNA (which by that stage was recognized to be the hereditary material of organisms) led to a resurrection of the idea in the 1960s.

The endosymbiotic hypothesis was fleshed out and popularized by Lynn Margulis. In her 1981 work *Symbiosis in Cell Evolution* she argued that eukaryotic cells originated as communities of interacting entities, including endosymbiotic spirochaetes that developed into eukaryotic flagella and cilia. This last idea has not received much acceptance, since flagella lack DNA and do not show ultrastructural similarities to prokaryotes. See also Evolution of flagella.

According to Margulis and Sagan, "Life did not take over the globe by combat, but by networking" (i.e., by cooperation).

The possibility that peroxisomes may have an endosymbiotic origin has also been considered, although they lack DNA. Christian de Duve proposed that they may have been the first endosymbionts, allowing cells to withstand growing amounts of free molecular oxygen in the Earth's atmosphere. However, it now appears that they may be formed *de novo*, contradicting the idea that they have a symbiotic origin.

Evidence

Evidence that mitochondria and plastids arose from ancient endosymbiosis of bacteria is as follows:

- Both mitochondria and plastids contain DNA that is different from that of the cell nucleus and that is similar to that of bacteria (in being circular and in its size).
- They are surrounded by two or more membranes, and the innermost of these shows differences in composition from the other membranes of the cell. The composition is like that of a prokaryotic cell membrane.
- New mitochondria and plastids are formed only through a process similar to binary fission. In some algae, such as *Euglena*, the plastids can be destroyed by certain chemicals or prolonged absence of light without otherwise affecting the cell. In such a case, the plastids will not regenerate.
- Much of the internal structure and biochemistry of plastids, for instance the presence of thylakoids and particular chlorophylls, is very similar to that of cyanobacteria. Phylogenetic estimates constructed with bacteria, plastids, and eukaryotic genomes also suggest that plastids are most closely related to cyanobacteria.
- DNA sequence analysis and phylogenetic estimates suggests that nuclear DNA contains genes that probably came from the plastid.
- Some proteins encoded in the nucleus are transported to the organelle, and both mitochondria and plastids have small genomes compared to bacteria. This is consistent with an increased dependence on the eukaryotic host after forming an endosymbiosis. Most genes on the organellar genomes have been lost or moved to the nucleus. Most genes needed for mitochondrial and plastid function are located in the nucleus. Many originate from the bacterial endosymbiont.

- Plastids are present in very different groups of protists, some of which are closely related to forms lacking plastids. This suggests that if chloroplasts originated *de novo*, they did so multiple times, in which case their close similarity to each other is difficult to explain. Many of these protists contain "secondary" plastids that have been acquired from other plastid-containing eukaryotes, not from cyanobacteria directly.
- Among the eukaryotes that acquired their plastids directly from bacteria (known as Primoplantae), the glaucophyte algae have chloroplasts that strongly resemble cyanobacteria. In particular, they have a peptidoglycan cell wall between their two membranes.
- These organelles' ribosomes are like those found in bacteria (70s).
- Proteins of organelle origin, like those of bacteria, use N-formylmethionine as the initiating amino acid.

Secondary Endosymbiosis

Primary endosymbiosis involves the engulfment of a bacterium by another free living organism. Secondary endosymbiosis occurs when the product of primary endosymbiosis is itself engulfed and retained by another free living eukaryote. Secondary endosymbiosis has occurred several times and has given rise to extremely diverse groups of algae and other eukaryotes. Some organisms can take opportunistic advantage of a similar process, where they engulf an alga and use the products of its photosynthesis, but once the prey item dies (or is lost) the host returns to a free living state. Obligate secondary endosymbionts become dependent on their organelles and are unable to survive in their absence.

The heterotrophic protist *Hatena* behaves like a predator until it ingests a green alga, which loses its flagella and cytoskeleton, while *Hatena*, now a host, switches to photosynthetic nutrition, gains the ability to move towards light and loses its feeding apparatus.

The process of secondary endosymbiosis left its evolutionary signature within the unique topography of plastid membranes. Secondary plastids are surrounded by three (in euglenophytes and some dinoflagellates) or four membranes (in haptophytes, heterokonts, cryptophytes, and chlorarachniophytes). The two additional membranes are thought to correspond to the plasma membrane of the engulfed alga and the phagosomal membrane of the host cell. The endosymbiotic acquisition of a eukaryote cell is represented in the cryptophytes; where the remnant nucleus of the red algal symbiont (the nucleomorph) is present between the two inner and two outer plastid membranes.

Despite the diversity of organisms containing plastids, the morphology, biochemistry, genomic organisation, and molecular phylogeny of plastid RNAs and proteins suggest a single origin of all extant plastids – although this theory is still debated.

6

Earlier Land Plants

INTRODUCTION

Beginning with this lab, we have tried to organize our systematic study of ancients plants by evolutionary lineage wherever possible. You should refer to the Virtual Paleobotany Laboratory for a composite cladogram with the current best interpretation of evolutionary relationships between extinct and extant plants discussed in this course. As you will see from that presentation, the early land plants discussed in this lab are completely paraphyletic. Why, then, do we discuss them together in a single lab? From the point of view of a botanist visiting the Silurian or Early Devonian, when these plants lived, the plants grouped together in this lab would represent the monophyletic crown group of embryophytes. Therefore, from a truly (Earth) historical perspective, we should consider these early plants together.

Among the earliest evidence of land plants (embryophytes) are, as you might expect, fossils of reproductive bodies (spores or cysts), because these are abundantly produced, highly transportable, and probably among the first plant structures to incorporate degradation-resistant chemical compounds (particularly sporopollenin). Such remains are recovered from the uppermost Ordovician and Lower Silurian siltstones and shales worldwide. If you are sampling marine rocks, however, it is difficult to establish a terrestrial origin for these fossils. The least equivocal microfossil evidence of early land plants would be trilete spores (which are not produced by algae), tracheids, plant cuticle (with stomata to distinguish it from some arthropod cuticle, which may look quite similar), and tetrahedral tetrads of spores. All of these structures are synapomorphic to the land plants (embryophytes) and so provide good evidence of the existence of this clade. However, we have few clues as to what these plants might have looked like or where they lived.

One important hint to the parents of the early spore tetrads comes from the cladogram itself. The fresh-water green alga *Coleochaete* , which is sister to the land plants, covers its zygote in sporopollenin. However, covering meiotic products (spores) with sporopollenin is a derived trait of the embryophytes. Surveying the next-derived embryophyte clades, we learn that some basal liverworts disperse their spores in tetrads, similar to those found in the Late Ordovician and Silurian. Therefore, a reasonable working hypothesis is that the early tetrads are produced by early bryophytes similar to liverworts, with the single dispersed spores that occur later being produced by later-evolving lineages such as hornworts, mosses, and vascular plants (tracheophytes).

SILURIAN-DEVONIAN TERRESTRIAL ALGAE AND PROBLEMATICA

Modern embryophytesembryophytes attest to the evolutionary proliferation of a single lineage of green algae that established a megascopic terrestrial flora. However, it is likely that the invasion of land was attempted by several groups, most of which were ultimately unsuccessful. We leave our phylogenetic framework for a moment to look at some of the early land plants that have boggled attempts at evolutionary classification.

Examine specimens of Parka, Prototaxites), and Protosalvinia (a.k.a. Foerstia). Some of these plants are thalloid much like early liverworts, hornworts and mosses might have been. Some have tubes, but their tubes are unlike those of any living organism. Reproductive structures are rare and confusing. However, it remains a paleontological truth that fossils belonging to entirely extinct lineages are the most difficult to understand.

The oldest demonstrably vascular plant is the Late Silurian genus Cooksonia, originally described from Wales. Cooksonia is characterized by small, slender axes that branch dichotomously. In fertile specimens, sporangia terminate each branch tip. Rooting structures or rhizomes have never been found, probably because most Cooksonia specimens are allochthonous, having been broken off and transported before burial. Cooksonia is only known from compression/impression fossils, so determining that it is a vascular plant is difficult. A few lucky specimens contain a suspicious dark trace in the center of the axis from which a few poorly-preserved conducting tubes have been isolated. However, other specimens lack this dark strand and it remains unclear whether this is a taphonomic artifact or whether some Cooksonia lacked conducting tissue. Cooksonia has also been discovered in eastern Europe and New York.

A grade is a group of organisms that have similar general organization but may or may not be related to one another. On first glance, the overall features of size and branching place all Cooksonia in a single group, a grade. However, closer evaluation of sporangial shape shows clearly that there is more variation within the genus than was previously suspected. Since reproductive structures like sporangia are generally better phylogenetic characters than are branching patterns, this variation gives us clues to the evolutionary relationships in the group. This analysis further suggests that the genus Cooksonia may include plants related to both the most primitive vascular land plants (Rhyniophytes) and to the lycopsids (Lab V). Consequently, the genus Cooksonia is not an evolutionary group (clade), but a mix of plants that bear some overall similarity (grade) and are near to the major branching in land plant evolution that sets the lycopsids off on their own evolutionary trajectory.

First, examine a hand specimen of the Rhynie Chert Chert is a finely crystalline silica that commonly forms in association with hot springs. Exactly how the Rhynie Chert of Scotland formed is still questionable, but it clearly represents an autochthonous deposition of plants in a swampy setting. Because of apparent rapid preservation by pulses of silica-bearing water, the Rhynie Chert preserves Early Devonian land plants in exquisite detail. How were these rocks formed? What type of preservation do they represent? What sort of anatomical and morphological features might you expect to see in the Rhynie plants? What features of the deposit suggest that the plant remains are an autochthonous (buried in place) rather than an allochthonous (transported to the site of burial)?

The Rhynie Chert preserves a variety of plants apparently growing together in an ancient community. Among them are Horneophyton, Aglaophyton, Rhynia and Asteroxylon. We will discuss the first three this week and save Asteroxylon for next week, when we discuss other members of its clade. Horneophyton, Aglaophyton, and Rhynia are all small plants, lacking leaves and roots, and bearing terminal sporangia on branching axes.

Aglaophyton

The plant formerly known as Rhynia major (Aglaophyton major) is approximately 18 cm in height and branches dichotomously. It appears to have a creeping rhizome. Sporangia are fusiform with the suggestion of a slight twist at the base.

Aglaophyton was removed from the genus Rhynia because the structures originally interpreted as tracheids in a central vascular strand, turned out on closer examination to be conducting tubes more like those of some mosses. Since unornamented conducting tubes would be plesiomorphic to the land plants, this character does not ally Aglaophyton with the moss lineage directly but does suggest a close sister-group relationship. Also, the main plant body of Aglaophyton is clearly a free-living sporophyte (sporangia are attached). The gametophyte of Aglaophyton appears to be Lyonophyton, another Rhynie Chert plant with naked axes and dichotomous branches; these branches terminate in antherida. Lyonophyton has some sort of conducting strand at the center of its axis, but the nature of the tubes (tracheids or no) is questionable. The two generations are linked together by shared epidermal features, which-as you remember from Laboratory III-carries intermediate weight of inference. If this plant truly has two independent free-living and morphologically similar generations, it takes an important place in plant life cycle evolution.

Horneophyton is a vascular plant with dichotomously-branched above-ground axes that terminate in sporangia. This plant differs from all other Rhynie Chert plants in having a corm-like base bearing numerous rhizoids. The vascular strand of Horneophyton becomes indistinct as it approaches the base of the plant and disappears altogether in the corm-like structure. Sporangia are cylindrical and branched, which also distinguishes this plant from all other Rhynie Chert taxa. The sporangia have a central column similar to that of some moss, but the clearly developed tracheids in the center of Horneophyton's axis (a derived state) remove it from the moss lineage. Furthermore, Horneophyton also appears to have a free-living gametophyte, Langiophyton . Although poorly known, this vascular plant was probably about 6 cm tall with cup-like terminal structures bearing archegonia. Horneophyton and Langiophyton are linked by epidermal features and the detail of conducting cells.

RHYNIA

Rhynia is also a vascular plant, somewhat monopodial branching to a height of about 18 cm. Upright axes branched from horizontal rhizomes that bore delicate rhizoids. Small bumps are noted along the axis, some of which may contain archegonia. *Rhynia gwynne-vaughanii* has a distinct protostele composed of distinct phloem and only a handful of xylem cells. The cortex is composed of parenchyma with abundant air spaces that resemble the spongy mesophyll of leaves and so may have functioned for photosynthesis. Sporangia are generally fusiform and located at branch tips. The presence of terminal sporangia and possible archegonia on the same plant have raised the possibility of a bryophyte-like life history. However, there is no structural evidence of a transition between sporophyte and gametophyte and the morphological relationship between the purported archegonia-bearing structures and the rest of the plant remain unclear. Further study will be needed to clarify *Rhynia's* contribution to our understanding of early land plant life cycle evolution.

The *Psilotum* Problem

Also in lab this week is the living plant *Psilotum*. *Psilotum* was originally classified in its own major group. Because of its very simple morphology (nearly dichotomous branching, protostele, lack of leaves and simple sporangia), many botanists believed that it is a survivor of the very primitive lineages of plants like those of the Rhynie Chert. Recent molecular analyses suggest strongly that *Psilotum* is allied with the most basal ferns. However, this conclusion is still debated on morphological grounds. If *Psilotum* is related most closely to ferns, several interesting questions emerge: What morphological homologies are possible between *Psilotum* and purported fern sister taxa? What are the developmental genetic changes that led to the stark changes in morphology? If the relationships suggested by the molecular phylogenies are correct, *Psilotum* may offer a tool for thinking about the morphological consequences of adding leaves to the simple axes of Early Devonian plants.

The excellent, in-situ preservation of the Rhynie Chert has offered us more than just a detailed look at the plants themselves. Through the Rhynie Chert, we see which plants were living together and get some idea of how many taxa grew together in a community. Since both sporophytes and gametophytes are present, some inference about life history evolution is possible. We also see other organisms that were living with the Rhynie Chert plants including a variety of arthropods. These small animals probably fed on decaying plant material although there is some suggestive evidence of animal attack on living plants. A fungus, *Paleomyces*, is also found infecting the soft tissues of the Rhynie Chert plants. It remains unclear whether *Paleomyces* was a decomposing organism, feeding on dead plant material, a parasite or a mycorrhizal symbiont. Some theoretical work suggests that mycorrhizal symbiosis might have helped early land plants solve the problems associated terrestrial life. However, many of these ideas await more detailed testing.

LUSH LIFE

What Early Land Plants Can Tell Us About Earth's Family Tree

The plant pioneers that moved from water to land had to change their way of life in order to adapt and survive. A team of biologists aims to find out which came first and how they evolved.

Half a billion years ago, the land on Earth was barren rock and ash. Only the water—the oceans, shallow seas, and lakes, teeming with simple aquatic plants, invertebrates, and primitive jawless fishes—held the secret of life.

Then somewhere, somehow, a brave new kind of green algae got a grip on the rocky shore. That momentous step, which may have happened in many places in various ways, made possible the world we know. We owe our existence to the land plants that evolved subsequently—to the oxygen they pump out, the nutrients they make and store, the soil they build, the animals they feed and shelter.

How did plants get that initial toehold on land, survive, and go on to diversify into the lush life we take for granted? A group of scientists dubbed "Deep Green" is dedicated to finding out.

"Deep," in this case, means deep in time. These biologists are fascinated with the "green" family because of its age and its central place in life on earth.

How did multicellular aquatic plants evolve? Which plants first colonized land? How are the early plant lineages related to each other? What genetic, cellular, and structural changes did they undergo?

With funding from the National Science Foundation's "Tree of Life" initiative, nine "Deep Green" scientists at SIUC and other institutions are tackling these questions to reconstruct in detail the family tree of the plant kingdom. The five-year, $3 million "genealogical" study is focusing on plant groups with an ancient heritage, such as algae, mosses, and ferns, where evolutionary relationships are fuzzy at best and where the most dramatic biological changes had to occur.

"If we can understand the relationship of living organisms—put together a 'big picture' of how life evolved—we can answer some important biological questions, which could influence everything from improving human health to managing the environment effectively," says SIUC plant biologist Karen Renzaglia, a member of the team.

Renzaglia, who will receive $375,000 from the NSF for her part of the research, studies cell-level changes in early land plants. "The constraints of living on land are pretty incredible," she says. "You have to figure out how to hold your body up against gravity, how to keep from drying out, how to move water and chemicals around in the body—and how to do sex out of the water!"

The earliest land plants to begin solving those problems are still the smallest: mosses, liverworts, and hornworts. Collectively called bryophytes, they're found on every continent but Antarctica. They hug the ground, absorbing water and nutrients directly through their cells. Skinny threads called rhizoids, precursors of root systems, anchor them to the soil.

Only after plants evolved true root systems and veins to ferry around water and nutrients could they grow very big. Giant tree ferns lifted their fronds up to the sun; rush-like horsetails and club mosses developed tall spore-bearing cones. These new types of plants, collectively called pteridophytes (Greek for "fern plants"), changed Earth's landscape dramatically.

Bryophytes and pteridophytes may have conquered the land, but they still rely on a film of water for reproduction. In mosses and their relatives, the plant produces sex organs containing sperm and egg cells. When the sperm are released, they "swim" to eggs on the same plant or nearby plants. Each fertilized egg sprouts a thin, shortlived stalk that bears spores along its sides or in a capsule at its tip. The spores fall to earth, where they germinate and produce new plants.

In ferns and their relatives, that sequence is reversed: The mature plant makes the spores. Turn over a fern frond and you'll see neat dots of spores bracketing the veins. When those spores fall to the soil, they each sprout a tiny plant—a mere speck on the forest floor or even underground—that exists to make egg and sperm cells. The sperm wiggle their way through water drops to fertilize nearby eggs, which then sprout new ferns.

Not until the evolution of seed-bearing plants, some millions of years later, did plant species free themselves from the need for water to get sperm and egg together. Seed plants—the trees, flowers, and grasses so familiar to us—package sperm in pollen, an ingenious solution that also enables them to reproduce at a distance.

Beyond these broad outlines, little is known for certain about early plant evolution.

"Major portions of this evolutionary tree are ambiguous," says Renzaglia. "There are so many gaps in our knowledge about the life cycles, structure, and development of early land plants. And almost nothing is known about the cellular details. These plants have evolved some fundamentally different mechanisms for [cell division and reproduction] that will give us information about evolutionary relationships."

Renzaglia studies the morphology (form and structure) and cell biology of plants. She has gained a worldwide reputation for her studies of sperm cells in bryophytes and pteridophytes. "Nobody knew the diversity of sperm in plants before we started doing this work," she says. Similarities and differences in these cells across species tell her a lot about when different groups of plants evolved and how they are related to each other.

Renzaglia suspects that the lowly hornwort, which coats itself with a water-holding layer of slime, was the earliest land plant that still exists today. In a separate project also funded by the NSF, she and biologist Joel Duff at the University of Akron are piecing together the family tree of the world's 150 or so kinds of hornworts.

So little is known about hornworts, she says, that "it's hard to know what defines a species or genus." Her morphological work, plus Duff's data on gene sequences and proteins that occur in these plants, are overturning current classification schemes.

"We never predicted the relationships we're coming up with," she says, "but they make sense in terms of which is the oldest hornwort and how the group developed."

Why sort out the hornworts? Why try to figure out the evolutionary pathways of the earliest land plants at all?

On an economic level, this information will help efforts in, for example, drug prospecting. If you identify one hornwort that produces a pharmaceutically useful compound, you'll want to zero in on that species' closest relatives so that you can analyze them for similar compounds.

On the environmental level, efforts to conserve biodiversity and save species will rely on knowledge about genetic relationships. Because we can't save everything, we should try to preserve the greatest genetic diversity, Renzaglia says. Species that are very distinctive genetically, with no close relatives, should take higher priority than species with abundant close relatives.

On a broader level, determining the history of early land plants will tell biologists more about how life forms function and have developed.

The "Deep Green" team has chosen for in-depth analysis some 50 species of plants that they believe represent major lineages among green algae, bryophytes, and pteridophytes. The team's leader, Charles O'Kelly of the Bigelow Laboratory for Ocean Sciences (Maine), is studying the morphology of the algae species. Renzaglia and her students are doing the same for the others.

To understand all the developmental stages of these plants, they're scrutinizing spores and spore capsules, egg and sperm cells, embryos, and mature plants. They're documenting dozens of characteristics, photographing specimens, comparing and compiling details previously reported about the species in the scientific literature, and archiving everything they find.

It's a huge amount of data. Take spores, for example. Renzaglia's team is recording, among other information, their shape and color, their "ornamentation" (whorls and ridges), their internal structure and development, and what types of molecules, such as proteins and sugars, are stored inside.

Using a scanning electron microscope, which bounces electrons off the surface of a sample, they're looking at plant structures such as stomata (pores) and sperm cells enlarged some 4,000 to 7,000 times—big enough to study the whip-like tails called flagella that enable the sperm to swim to their destination. Using a transmission electron microscope, which shoots electrons through a thin section of tissue, they're viewing the inner workings of cell nuclei and chloroplasts (the cell bodies that carry out photosynthesis). This work is being done under the aegis of SIUC's IMAGE facility.

Two tiny bryophyte species whose microanatomy differs considerably can look virtually identical to the naked eye. Conversely, one species can vary considerably in its outward appearance depending on environmental conditions. So cellular features will be key to learning how these early plants evolved.

For example, some hornworts have big chloroplasts that look identical to those in green algae, Renzaglia and Duff have found, while other hornworts have smaller chloroplasts but more of them. The latter species are farther away, evolutionarily, from the algae/hornwort split.

As an undergraduate in Renzaglia's lab, Kelly Wood studied plant sperm exposed to a mutation-causing chemical that damaged their flagella. As a result, these cells had trouble swimming. Fast forward: Wood went on to medical school, where she learned about human reproductive and respiratory cells with damaged cilia (hairs that move the cells themselves or surrounding material). She discovered that these cilia had the same kinds of damage as the plant flagella, which are structurally similar. Medical researchers now have a new experimental model that may shed light on the nature of the human disorder.

Intriguing information also will come from genetic studies of early land plants. Other team members are analyzing DNA from the same plant specimens that Renzaglia and O'Kelly are working with. Among other things, they are creating "reference libraries" of thousands of chromosome snippets from each specimen for cross-species comparison. The more genes that two organisms have in common, the more likely it is that they fall on the same branch or twig of the evolutionary tree. "If we understand these gene sequences [from early plants], it will help tell us how genomes change over time," Renzaglia says.

The genetic analyses will generate gigabyte after gigabyte of valuable data to be combined with the morphological data. Both are needed.

"Genes don't tell you what the plant looks like or how it functions," Renzaglia says. "And without information on morphology, you won't know what structural changes have taken place during evolution." Gene studies will help morphologists focus their efforts, and vice-versa. Although scientists ultimately will tease out the function of many genes, particularly those that are shared by many species, such analysis is time-consuming, and it can't be done for every gene in every kind of organism.

The fact that many species and even entire lineages have gone extinct over millions of years also complicates genetic analysis. "It's possible that all of the early hornworts died off except for one branch, which persisted unchanged for a long time and then underwent recent speciation," Renzaglia speculates by way of example. "Then the group would be old, but much of the genetic diversity would be recent. In colonizing land, probably thousands of successful species died off. It's frustrating—we only have bits and pieces of evidence."

Why not just study fossil plants? "You can't get DNA out of fossils," Renzaglia notes. "You can't see development; you can't see sperm. You can't, in many cases, see cellular details. And we don't have a good fossil record of the bryophytes—they're small and fragile."

Pinning down the details of plant evolution will take a grand synthesis of fossil, genetic, and morphological evidence. And that poses the team's biggest technical challenge: how to wade through all that data. The team will develop new methods of organizing, combining, and analyzing diverse data sets to draw conclusions about evolutionary history. Their Holy Grail, and a major goal of the NSF initiative, is an analytical model that can be applied to determine the evolutionary relationships of other organisms.

Besides Renzaglia and O'Kelly, scientists at the University of California - Berkeley, the University of Washington, Yale University, Utah State University, and Lawrence Berkeley National Laboratory are working on the project.

Renzaglia is passionate about this work. "It has the potential to uncover all kinds of new information," she says. "There's biodiversity at every level of life, including the cellular level. research has opened up a world nobody's known about.

7

Evolution and Diversity

INTRODUCTION

By human standards, 600 million years is a long, long time. Our human species is less than 2 million years old, birds and mammals have existed for no more than 200 million years, dinosaurs first appeared 230 million years ago, and the most primitive land plants arose about 425 million years ago.

But in order to observe the first arthropods, (phylum Arthropoda) we would have to travel back even further in time — to a point nearly 600 million years ago (the Cambrian period of the Paleozoic era) when bacteria and marine algae were the dominant forms of plant life, small invertebrate animals were abundant in warm, shallow seas, and land masses were still largely devoid of life.

It seems incredible that we can even think of studying events that happened so long ago. But biologists and paleontologists have a wide array of tools and techniques that allow them to probe the world of today for clues that illuminate events of our distant past. Such clues form an ever-growing patchwork of data that has begun to merge into a rough outline of the arthropods' phylogenetic tree. Although most of our hypotheses about evolutionary beginnings are tentative and controversial, they provide a valuable framework in which to study similarities, differences, and relationships among all surviving taxa. Our understanding of each phylogenetic group is enhanced by learning how it has been shaped by the selective pressures of the past, and how it differs from its nearest relatives of the present.

Tools and Techniques

The science of paleontology (the study of prehistoric forms of life) and the principles of systematics (classification of organisms based on structural or evolutionary relationships) are the traditional foundations of phylogenetic research. Both endeavours are based on two important assumptions:

All forms of life share a similar DNA-based system of inheritance; and

The process of natural selection has remained virtually unchanged throughout evolutionary time.

The fossil record is the pre-eminent database for phylogenetic research. Prehistoric arthropods were generally small and delicate. They did not preserve as well as larger animals with teeth, bones, or shells, but are more abundant as fossils than other soft-bodied invertebrates because their rigid exoskeleton maintained its shape and did not decay rapidly after death. Fossil arthropods can be found in sedimentary rock strata and in deposits of coal, shale, or volcanic ash throughout the world. Good fossil beds containing arthropods are relatively scarce, so the fossil record tends to be patchy. Many species are known to science by only a single specimen, while others are recognizable as species that still live today.

Once entangled in the sticky resin, captive insects were preserved in every detail as the viscous sap gradually hardened to glass-like consistency. Amber deposits, containing embedded insects from the early Cretaceous period through recent times, can be found in northern Europe (around the shores of the Baltic Sea), in Asia (e.g. Myanmar and Siberia), and in North America (Canada, Alaska, Mexico, and the Dominican Republic).

Until recently, scientists were forced to rely almost exclusively upon their own interpretations of the fossil record to reconstruct phylogenetic relationships. Many structural similarities between organisms are regarded as evidence of homology (kinship), while many dissimilarities signify unrelatedness. Problems often arise because of differences in interpretation, absence of intermediate forms, loss of specialized adaptations (reversion to more primitive conditions), and development of similar characteristics in unrelated groups (convergent evolution). Classical taxonomists have always been constrained by the overwhelming volume of data and the subjectivity of the process.

Within the past 30 years, however, exciting developments in chemistry, physics, and computer science have given evolutionary biologists a number of powerful new tools that furnish more precise and objective information about fossils and their relationship to extant life-forms.

The approximate age of a fossil is vital information in assessing where it might fit into the family tree of its relatives. Fossils traditionally have been dated by the rock strata in which they occur. Relatively accurate ages of these strata can be inferred by measuring the concentration of a radioactive isotope (such as carbon-14) or by determining the ratio between a radioactive element (uranium-235, thorium-232, or potassium-40) and its spontaneous decay product. Plants and animals assimilate carbon-14 from the atmosphere only while they are alive, so it is possible to determine approximate age of a fossil by measuring how much carbon-14 has not yet decayed to carbon-12. Carbon dating is reliable only on specimens young enough to retain measurable amounts of carbon-14 (less than 50,000 years old).

Older strata of rock that bear trace amounts of radioactivity may be dated by measuring the ratio between an unstable isotope (such as uranium-235, thorium-232, or potassium-40) and its stable decay product (lead-207, lead-208, or argon-40 respectively). The very long half-life of these isotopes (from 713 million to 14 billion years), makes them potentially useful for dating samples that range in age from 100,000 years to well before the beginning of prehistoric life.

Digital computer technology provides a new and highly objective approach to evaluating similarities and differences among taxa. Statistical techniques, such as principal components analysis, cluster analysis, factor analysis, multidimensional scaling, and discriminant analysis can be applied methodically to large data sets containing morphological, behavioral, or biochemical information about taxonomic groups of interest.

Phylogenetic Trees

The calculations may employ both "primitive" and "advanced" character states (the phenetic approach) or rely exclusively on "advanced" characteristics (the cladistic approach). In either case, the resulting "phylogenetic trees", variously known as phenograms, cladograms, or dendrograms, are intended as visual representations of perceived biological affinity. They may (or may not) depict true evolutionary relationships.

Biochemistry

Enzymes and metabolic pathways can reveal inherent patterns in nutrient processing, chemical defense, locomotion, intercellular and intracellular communication, or homeostatic mechanisms. Similarities and/or differences in these basic life processes are often consistent within an evolutionary lineage. The sudden appearance of a novel biochemical mutation may confer selective advantages and eventually lead to adaptive radiation of new species from a common ancestor.

Automated methods of sequencing nucleotides in DNA (and ribosomal RNA) have opened a new window to the past. We are discovering that the evolutionary "distance" between two organisms can be inferred from the number and frequency of changes in the sequence of base-pairs in corresponding regions of DNA (or RNA). By assuming that mutation rates have been relatively constant throughout geological history, it is possible to estimate how long it has been since any two groups diverged from a common ancestor. Recently, biologists have even extracted and sequenced DNA from fossils. This amazing feat was accomplished by isolating fragments of DNA from insects (termites and stingless bees) preserved in amber. The fossil DNA was cloned (replicated) using polymerase chain reaction (PCR), an enzymatic method that mass-produces identical copies of a DNA molecule. When enough of the prehistoric DNA had been manufactured, it was sequenced and compared to similar fragments from the insects' living relatives. If these techniques prove applicable to a broad range of preserved specimens, they will allow us to study the genetic makeup of extinct organisms, and revolutionize our ability to explore relationships among the arthropods of today and their long-extinct ancestors.

THE FIRST ARTHROPODS

A variety of marine worms (Annelida and Protoannelida) lived in the ocean sediments during the Cambrian period. These creatures were bilaterally symmetrical, soft-bodied, and multisegmented. They had no distinct head capsule and lacked both eyes and antennae. Some species may have had lobe-like lateral appendages similar to the parapodia of polychaete worms that thrive in the muddy sediments of today's ocean floor. But structural differences between the lateral appendages of polychaetes and early arthropods suggest that these two groups diverged from a common annelid ancestor some 500-600 million years ago. All annelids were soft-bodied. They decayed rapidly after death and were not well-preserved in the fossil record.

The Onychophora or velvet worms were suggested as a possible evolutionary link (intermediate) between the annelids and the arthropods. These mostly-tropical invertebrates, with only 75 extant species, live on the forest floor among moist, decaying leaves and feed on a mixed diet of plant and animal tissues. Like annelid worms, the Onychophorans have segmented bodies containing paired excretory organs (nephridia) and a combination of both male and female sex organs (monoecious). They lack a true exoskeleton, but the body is covered by a soft, chitinous cuticle. In true arthropod fashion, this cuticle is periodically shed (molted) to allow for continued growth. Onychophorans also share numerous other characteristics with the arthropods, including antennae, an open circulatory system, walking legs with claws, paired mandibles, and a system of slender air tubes (tracheae) for respiration.

Velvet worms may or may not prove to be direct ancestors of the arthropods, but regardless of classification, it seems reasonable to assume that the very first arthropods were distinctly worm-like in structure and appearance. They were also different in at least one important way: they had an exoskeleton. This structure, secreted by epidermal cells, was more rigid than annelid cuticle and necessitated membranous "joints" to provide the flexibility needed for movement. This exoskeleton (and its jointed appendages) is the hallmark of the arthropods.

Arthropod Evolution

A classical controversy that still rages among invertebrate zoologists, including the entomologists, relates to the evolutionary pathway (or pathways) of adaptive radiation that arthropods followed as they gradually diverged from primitive ancestors. The traditional, and certainly more conservative approach, assumes that arthropods arose only once from protoannelid ancestors. This monophyletic argument is based on the observation that many features, such as the exoskeleton, open circulatory system, hemocoel, etc., are shared by nearly all taxa within the group and appear to be homologous (i.e. have the same evolutionary origin). A contrary view is taken by other biologists who argue that arthropod-like organisms must have evolved more than once (perhaps as many as four times) in geological history. Support for this polyphyletic approach is found in embryological development and in comparative studies of the mouthparts and other appendages. Proponents of the polyphyletic hypothesis claim that many similarities between taxa have arisen coincidentally, through the process of evolutionary convergence.

The Monophyletic Hypothesis

Ancestors in several important ways: they have a chitinous exoskeleton, jointed/segmented appendages, a well-developed head and mouthparts, striated muscles, and an open circulatory system with a dorsal heart. Unlike annelids, the arthropods produce large, yolk-laden eggs that are encased in a proteinaceous shell. *Ciliated nephridia*, paired segmental excretory organs found in annelids and onychophorans, have been replaced in arthropods by specialized excretory organs located on the head (green glands in crustacea), near the legs (coxal glands in horseshoe crabs), or in the abdomen (Malpighian tubules in terrestrial arthropods). Remarkable similarities in the chemical composition of the exoskeleton and in the ultrastructure of the compound eyes in organisms as diverse as millipedes, shrimp, and horseshoe crabs provide strong evidence that all of these groups (myriapods, crustaceans, and chelicerates) must have evolved from a common ancestor. The monophyletic classification scheme, proposed by Boudreaux recognizes three evolutionary lineages within the phylum Arthropoda: Trilobita, Chelicerata, and Mandibulata. The trilobites became extinct by the end of the Permian Era and therefore represent an evolutionary "dead end". Chelicerata includes all arthropods with fang-like mouthparts (chelicerae): spiders, ticks, and mites (Arachnida), horseshoe crabs (Xiphosura), and sea spiders (Pycnogonida). Mandibulata includes all arthropods that have chewing mouthparts (mandibles): crustacea, myriapods, and insects. Early in the Paleozoic Era, the mandibulate lineage divided into at least one group that continued a marine lifestyle (the crustacea), and another group that adopted a terrestrial lifestyle. This terrestrial lineage, which encompasses all present-day myriapods and insects, is known as the superclass Atelocerata, a taxon first described by Heymons in 1901.

Boudreaux's classification scheme excludes onychophorans because they lack a true exoskeleton, but other workers disagree. The Onychophora as closely related to myriapods. In their monophyletic classification scheme , Crustacea and Chelicerata are the most primitive groups. Insects, myriapods, and onychophora are grouped together in the superclass Uniramia because of similarities in leg structure and locomotion.

A great deal of controversy surrounds the evolution of arthropod legs and wings. Proponents of monophyletic classification argue that the legs of all arthropods are homologous (have a common evolutionary origin); opponents claim there is too much diversity in leg structure to justify a single ancestor. Jarmila Kukalová-Peck, a Canadian entomologist, has proposed that "primitive" arthropods may have had as many as eleven segments in each walking leg. She cites evidence from an extinct fossil insect to support her claim that the eight-segmented legs of spiders, the seven- to nine-segmented legs of crustacea, and the five-segmented legs of insects all exhibit some degree of reduction from the "primitive" groundplan.

The Polyphyletic Hypothesis

If we embrace the idea of polyphyletic origins, then arthropods are represented by as many as four major phyla—each of which is presumed to have evolved separately from primitive annelid ancestors:

Trilobita—4,000+ species—including all trilobites (extinct marine organisms that were abundant during the Paleozoic era.)

Chelicerata—70,000 species—including spiders, scorpions, mites, ticks, horseshoe crabs, and sea spiders.

Crustacea—30,000 species—including shrimp, crabs, lobsters, woodlice, barnacles, amphipods, branchiopods, and copepods.

Uniramia—1.2 million species—including onychophora, millipedes, centipedes, pauropods, symphylans, and insects.

Evidence to support the polyphyletic hypothesis can be found in the comparative anatomy of appendages and in the embryonic development of the head and mouthparts. Charlotte Manton, one of the founders of the polyphyletic hypothesis, suggested that there is a fundamental difference between the appendages of crustacea and those of other arthropods such as insects and myriapods (millipedes and centipedes). Manton argued that crustacean appendages are biramous; that is, two apical units (rami) are attached to a single basal unit. Appendages of other arthropods are uniramous: a single apical segment is attached to a single basal segment. Manton believed that crustaceans evolved from annelid worms similar to marine polychaetes of today, and that all other arthropods evolved from annelid worms that were more similar to the onychophora. This hypothesis is also supported by D.T. Anderson whose studies of arthropod eggs has revealed that initial cell division in crustacean embryos is holoblastic (spiral cleavage), whereas the eggs of all other arthropods are meroblastic (superficial cleavage). The eggs of all known annelids are holoblastic.

The Myriapods

Regardless of whether we adopt a monophyletic or a polyphyletic classification scheme, it is apparent that the insects have more in common with the myriapods than with any other taxon of the Arthropoda. Insects and myriapods are adapted to a terrestrial lifestyle, exhibit similar embryological development, and share many similarities in the morphology of the head and mouthparts.

Chilopoda (centipedes) and Diplopoda (millipedes) as well as two other classes that are not as well-known: Symphyla and Pauropoda. All of these organisms live in moist environments near the soil surface (e.g., under stones, amid leaf mold, in rotting wood, etc.). Centipedes are primarily predators, the others are scavengers and herbivores.

All myriapods exhibit ametabolous development (there is no significant change in body form as they mature). Eggs hatch into immatures (called young) that are similar to adults in most respects except size and sexual maturity. Like all other arthropods, they grow by molting, but they may also increase in length by adding an additional body segment at each molt. This type of growth (called anamorphosis) is common in myriapods, but it occurs only rarely in more "advanced" arthropods and is usually regarded as a "primitive" characteristic (pleisiomorphic condition).

Functional body regions: a head and a trunk. The head is specialized for sensing the environment (eyes and antennae) and ingesting food. The trunk is adapted for locomotion and also contains most of the internal organs. Pauropods and symphylans are generally small (under 5 mm in length) with little or no body pigmentation — they are frequently creamy white in colour. Symphylans have beaded antennae and 10 to 12 pairs of legs. Pauropods have branched antennae, nine pairs of legs, and plates of exoskeleton on the dorsal surface that usually span more than one body segment. Millipedes and centipedes are usually larger (more than 5 mm in length) and dark brown to black (although some species may be quite colourful). Centipedes have at least 15 pairs of legs — the first pair, immediately behind the head, contain poison glands and are modified as fangs for killing prey. Millipedes have at least 30 pairs of legs, two pair per body segment. They are not as fast-moving as centipedes and often curl into a ball or release a defensive spray as protection from predators.

Most biologists and paleontologists feel that there is sufficient morphological, embryological, and fossil evidence to justify a claim that the basic structural plan of the insect head evolved very early in the myriapod lineage. As the head assumed a more prominent role in the organism's survival, it increased in size and usefulness by assimilating adjacent trunk or body segments that previously had been-adapted for locomotion. In this process of cephalization, the leg appendages evolved into mouthparts that became adapted for catching, processing, or manipulating food items. These mouthparts are said to be segmentally homologous with walking legs. Indeed, if this is the case, we might expect to find "primitive" organisms that have only mandibles (monognathous), "less primitive" organisms having both mandibles and maxillae (dignathous), and "advanced" organisms with mandibles, first maxillae, and second maxillae (trignathous). In fact, monognathous forms have never been found (living or fossilized), but the dignathous condition occurs in Pauropoda and Diplopoda, and trignathous mouthparts are found in centipedes, symphylans, and all hexapods (including insects and their close relatives).

The symphylans and the hexapods (but not the centipedes) have experienced an anatomical fusion of the second maxillae. This medial fusion of the left and right side appendage has produced a mouthpart called the labium. It is tempting to interpret this structural similarity as evidence of a close evolutionary affinity between Symphyla and Insecta. In fact, this was the basis for the "symphylan theory" of insect evolution that was proposed by Calman and Imms in 1936. Despite its intuitive appeal, the "symphylan theory" is not widely accepted today because of other, more extensive differences in leg and abdominal structure between these two groups.

THE HEXAPODS

Judging from the fossil record, hexapods must have diverged from myriapodan.

APTERYGOTE INSECTS

The insects probably evolved from some primitive member of the superclass Myriapoda during the Silurian period (400-440 million years ago). At this time in geological history, vascular plants were just beginning to evolve and invade the dry land. These plants undoubtedly represented a new and largely unexploited source of both food and shelter, and it is apparent from the early fossil record that insects were quick to take advantage of this new resource. Their newfound success in these harsh, terrestrial environments was certainly due, in part, to an exoskeleton with an outer layer of wax to minimize desiccation.

Although there are no fossil remains of the first insects, it is reasonable to speculate from structures of the head and mouthparts that insects evolved independently from other hexapods. As mentioned previously, all primitive insects have ectognathous mouthparts (mandibles and maxillae not enclosed within a head cavity). These mouthparts are generally directed downward from the ventral side of the head (hypognathous) rather than forward from the front of the head (prognathous) as in other hexapods.

Insects also evolved with an additional sclerite (the clypeus) on the anterior surface of the head capsule. Muscles extend from the clypeus to the upper walls of a pre-oral cavity (the cibarium) behind the mouthparts. When these muscles contract, the cibarium enlarges (like opening a bellows) and sucks food toward the mouth.

All insects are epimorphic (the number of body segments does not change over time). The primitive body plan includes a head, a three-segmented thorax, and an eleven-segmented abdomen. Paired abdominal appendages (slender styles or short pegs) can be found in only the most primitive insects. These appendages are probably remnants of myriapodan legs. The external opening of an insect's reproductive system, the gonopore, never occurs on the last abdominal segment (as in other hexapods). In male insects, the gonopore is located on the ventral surface of the ninth abdominal segment. In females, it is usually on or behind the eighth abdominal segment.

Insects are also characterized by a distinctive antennal structure (absence of muscles beyond the two basal segments), a wax layer covering the exoskeleton, and a complete respiratory (tracheal) system with external valves (spiracles) that can be opened to allow gas exchange or closed to prevent desiccation.

The earliest fossil insects are found in Devonian rock. They are strikingly similar to insects that we recognize today as bristletails. These "most primitive" of all insects are members of the subclass apterygota (from the Greek "a-" meaning without and "pterygo" meaning wing). They are completely wingless and have ametabolous development. In the past, entomologists grouped all primitive apterygote insects into a single order (Thysanura). More recently, however, the trend has been to divide this subclass into three orders: Archaeognatha, Thysanura, and Monura. Of these, only Archaeognatha and Thysanura have survived to the present time.

These apterygote orders share several "primitive" characteristics with the myriapods and non-insect hexapods. First, they continue to molt even as adults. Some Thysanura, for example, may live for several years and molt 40-50 times. Second, apterygote insects have short, segmented appendages along the sides of the abdomen. These structures appear to be homologous with the walking legs of myriapods. And third, males produce spermatophores that are laid on the ground and later picked up by the female (external fertilization). Some species have developed elaborate courtship rituals that insure rapid and efficient exchange of the spermatophore.

Archaeognatha

The order Archaeognatha (sometimes called Microcoryphia) includes insects commonly known as jumping bristletails. The members of this order are distinctive because their mandibles connect with the head capsule in only one place (monocondylic). This single point of articulation allows an auger-like vertical movement of the mandibles that is effective for removing lichens and algae from the substrate. Monocondylic mandibles are regarded as an "ancestral" characteristic (pleisiomorphic condition) because the mandibles of all other insects, Thysanura included, have two points of articulation with the head (dicondylic). Other ancestral features of Archaeognatha include external fertilization, claws on the maxillary palps, segmented abdominal stylets, and a ring-like segment (subcoxa) at the base of each leg.

Most bristletails live in grassy or wooded habitats where they are most likely to be found in leaf litter, under bark, among stones, or near the upper tidal line in coastal areas. They are most active at night, feeding as herbivores or scavengers on algae, mosses, lichens, or decaying organic matter.

Sexual maturity is reached after at least eight juvenile instars spanning up to two years. Molting continues periodically even after adulthood. The sexes are separate, but copulation does not occur.

Archaeognatha

Males produce a packet of sperm (spermatophore) and leave it where a female is likely to find it. Females cannot store sperm (they lack a spermatheca), and evidently acquire a new spermatophore before each bout of egg laying. Eggs are laid singly or in small groups (<30). Some species have elaborate courtship rituals to insure that females are able to locate a spermatophore.

Thysanura

The order Thysanura (sometimes called Zygentoma) includes the insects commonly known as silverfish and firebrats. Fossils of this order first appear in rock from the early Carboniferous period. Features of the head and mouthparts are more specialized than those of Archaeognatha. Dicondylic mandibles move only in a transverse direction allowing food to be ground between them on the molar surface.

Silverfish are fast-running insects that hide under stones or leaves during the day and emerge after dark to search for food. A few species are resistant to desiccation and well-adapted to survive in domestic environments such as basements and attics. Silverfish are scavengers or browsers. They survive on a wide range of food, but seem to prefer a diet of algae, lichens, or starchy vegetable matter.

Thysanurans may be rather long-lived—three years is probably typical and up to seven or eight years may be possible. They continue to molt frequently, even after reaching adulthood.

Silverfish have an elaborate courtship ritual to insure exchange of sperm. The male spins a silken.

Thysanura

Thread between the substrate and a vertical object. He deposits a sperm packet (spermatophore) beneath this thread and then coaxes a female to walk under the thread. When her cerci contact the silk thread, she picks up the spermatophore with her genital opening. Sperm are released into her reproductive system, and then she ejects the empty spermatophore and eats it.

Subclass Pterygota

About the same time primitive insects began to diversify and exploit terrestrial habitats, other types of arthropods (especially spiders and scorpions) were also invading the land. Almost all of these arachnids were predatory, and it seems likely that insects were a major component of their diet. If this is true, then the selective pressure of predation may explain why most surviving members of the Apterygota are cryptozoic (adapted for life in concealed places—in leaf litter, under bark, etc.)

Predation also may have been the driving force behind an event that is probably the single most significant landmark in the evolutionary history of insects—development of wings. Flight was not only the ultimate escape from earth-bound predators, but it also proved to be a rapid form of transportation, an efficient mode of dispersal, and a convenient way to find a mate!! Insects were the first living organisms to master flight, and for more than 150 million years (until the first flying reptiles appeared in the Jurassic period) the insects were undisputed rulers of the air.

How, when, and where did wings evolve? These questions have puzzled entomologists since the days of Charles Darwin. The fossil record is not very informative because fully winged species appeared rather suddenly in the Carboniferous period (about 380 million years ago) with no clear evidence of ancestral precursors. Like all other adaptations, the development of functional wings must have been influenced by the process of natural selection acting over a span of many generations. But what adaptive value could there have been for protowings that were too small for sustained flight?

Some biologists have speculated that wings may have developed as modifications of tracheal gills in aquatic insects (the gill theory), from flaps of integument covering thoracic spiracles (the spiracular flap theory), from pre-existing appendages near the legs (the endite-exite theory), or as lateral extensions of dorsal plates on the thorax (the paranotal lobe theory). These structures were not necessarily related to flight—they may have played a role in sexual selection (helping to attract a mate), in thermal regulation (absorbing heat energy like a passive solar panel), or in aerodynamics (improving the stability of a jumping or gliding insect). Any of these theories could account for the development of large wing-like structures, but none have been widely embraced because they all fail to explain how or why such appendages acquired a hinge-joint mechanism, strong thoracic musculature, and the neural complexity necessary for sustained flight. The debate continues, but so far it has produce more heat than light.

In 1994, James Marden and Melissa Kramer proposed an attractive new theory that may add fresh insight to the puzzle of wing evolution. They suggested flight may have originated among aquatic insects living on the water surface. Such insects could have used their gills or gill covers as oars to row across the water. Selective pressure for a more efficient rowing mechanism could lead to larger oars (protowings) and stronger muscles. High speed rowing movements might easily progress into skimming along the water surface, and eventually to self-powered flight. In fact, surface-skimming behavior has been observed in both mayflies (Ephemeroptera) and stoneflies (Plecoptera).

There is direct experimental evidence of a correlation between wing size (surface area) and skimming velocity in *Taeniopteryx burksi*, a small winter stonefly with a feeble mastery of flight. Marden and Kramer's theory is noteworthy because it accounts for the progressive development of wings as well as the muscles needed to power them.

Based on the comparative morphology of wing structure, venation, and flight mechanics, it appears that the ability to fly evolved only once in the class Insecta. All remaining orders are grouped into a single subclass, the Pterygota (from the Greek word "*pterygo*" meaning a wing) because they are believed to have all descended from winged ancestors. Some pterygote insects (lice and fleas, for example) have acquired a specialized lifestyle where wings have been lost because they provided no selective advantage. These insects are said to be "secondarily wingless"—their winglessness is a derived adaptation (apomorphic condition) not to be confused with the primitively wingless state (pleisiomorphic condition) of apterygote insects.

Nymph with wing pads Concurrent with the development of wings, the first pterygote insects also show gradual developmental changes in body form as they mature. Wings and external genitalia grow and develop during the immature stages, becoming complete and functional only after the final molt into adulthood. This gradual change in body form is called hemimetabolous development (incomplete metamorphosis). Immature stages of hemimetabolous insects are known as nymphs (or naiads if they are aquatic) rather than young (as in ametabolous insects).

For descriptive purposes, all insects with hemimetabolous development are grouped together as Exopterygota. This name (from the Greek "*exo*" meaning outer and "*pterygo*" meaning wing) reflects the fact that wings develop externally as outgrowths of the nymph's body wall. They grow larger with each molt and usually become visible as stationary "wing pads" on the top or sides of the thorax. Functional wings occur only in adult insects.

The Paleopterous Orders

The most primitive winged, hemimetabolous insects have large, membranous wings that are held continuously out to the side of the body or over the back of the thorax. These insects do not have the ability to fold their wings down flat over the abdomen. All orders that share this "primitive" condition are grouped as Paleoptera, a subdivision (infraclass) within the subclass Pterygota. The word paleoptera is derived from the Greek words "*paleo-*" meaning primitive, and "*ptera*" meaning wings.

Based largely on the fossil record, entomologists now recognize up to seven orders of paleopterous insects. Most of these orders were relatively abundant during the

Carboniferous period, but they disappeared suddenly in the Permian (about 250 million years ago). The largest insect that ever lived, *Meganeura monyi* Brongniart, was a member of one of these extinct orders, the Protodonata. It was a damselfly-like insect with a body length of 30 cm (1 foot) and a wingspan of nearly 75 cm (2.5 feet).

Of all the paleopterous orders, only two (Ephemeroptera and Odonata) have survived to the present time. Immatures of both orders are exclusively aquatic (naiads = water nymphs). Adults are predominantly aerial: their legs are often reduced or adapted for clinging or grasping, never for walking or running. Adaptations for internal fertilization (penis-like structures) first appeared in adults of the Paleoptera.

Ephemeroptera

The Ephemeroptera (or Ephemerida) are called mayflies, but may also be known locally as shadflies, lakeflies, or spinners. Immatures are aquatic; they generally live in unpolluted habitats with fresh, flowing water. Some species are active swimmers, others are flattened and cling to the underside of stones, a few are burrowers who dig U-shaped tunnels in the sand or mud. Most species are herbivorous. Their diet consists primarily of algae and other aquatic plant life scavenged from surrounding habitat. Some species mature quickly, in as little as four weeks, while others develop more slowly (one to four years per generation).

Once a mayfly completes development as a naiad, it leaves the aquatic environment, often rising to the water surface in a bubble of air. It quickly molts to a winged form (the subimago) and flies to a nearby leaf or stem. The subimago is a transitional stage. Within a few hours, it molts again into an imago, a sexually mature adult. The imago usually has transparent wings and a smooth, shiny exoskeleton in contrast to the cloudy wings and dull, pubescent body of the subimago. Mayflies are the only insects that molt again after they have wings.

Most adults are delicate insects with a very short lifespan. They do not feed (mouthparts are vestigal), and some species emerge, reproduce, and die in a single day. Males generally fly in swarms that undulate in the air 5-15 meters above the ground. Females fly into the swarm and are quickly grabbed by a male. Copulation takes place in flight, and the female usually lays her clutch of eggs within minutes or hours. Males die shortly after mating, females usually die soon after oviposition.

Ephemeroptera

Odonata, dragonflies and damselflies, are predaceous both as immatures and adults. The adults are quick, agile fliers that are generally considered beneficial because they feed on large numbers of small, flying insects like gnats and mosquitoes. Legs are used either as a basket for catching prey or as grapples for clinging to emergent vegetation. Eggs are laid singly in fresh water; females often hover over open water and dip their abdomen as they oviposit.

Eggs hatch into aquatic immatures (naiads) that feed opportunistically on other forms of aquatic life including mayfly naiads, small crustaceans, annelids, and molluscs. Some of the larger dragonfly naiads will even attack small fish and tadpoles. All immature odonata have a specialized labium for catching prey. Folded under the head and thorax when not in use, the labium can be extended rapidly toward potential prey. Hooked lobes at the tip of the labium grasp or impale the prey and draw it back to the mouth as the labium retracts.

Damselfly naiads are usually more slender than dragonfly naiads, and have three leaf-like gills at the end of the abdomen. Dragonfly gills are located internally, within the rectum, where bellows-like contractions of the rectal muscles cause oxygenated water to circulate in and out.

The Neopterous Orders

The large, membranous wings of the Paleoptera were not always an advàntage. The insects had to land occasionally to rest, feed, or mate, and once on the ground, big bulky wings would only attract the attention of predators and hinder movement through dense foliage. If the wings became broken or shredded by objects in the environment, they would be a major liability. It is probably worth noting that Ephemeroptera and Odonata, the only surviving members of the Paleoptera, do not live on the ground. Wings are their only means of locomotion; legs are used only for grasping or clinging. The adults of both orders have an aerial lifestyle that puts them beyond the reach of earth-bound predators.

It seems likely, therefore, that selective pressures on the first winged insects heavily favored the development of some mechanism for folding the wings against the body after landing, making them less conspicuous, less awkward, and less susceptible to breakage. Such an adaptation first appeared 350-400 million years ago (in the late Devonian or early Carboniferous period). The progeny of these wing-folders are placed in the infraclass Neoptera, meaning "new wing". These insects represent a remarkably successful lineage; they became the ancestors of all "higher" orders of insects.

The "new wing" was articulated at its base so it could be folded flat over the insect's abdomen when not in use. The folding mechanism is facilitated by an elastic hinge and a single flexor muscle attached to a small plate (third axillary sclerite) at the base of the wing. With this adaptation, winged insects could be as agile on the ground as they were in the air. Relative freedom from predation plus an abundant food supply (in the form of green plants) gave these winged insects a unique opportunity to exploit new ecological niches. Indeed, there was an unprecedented explosion of species during the upper Carboniferous period — at least eleven new insect orders appeared within a span of less than 30 million years.

During this burst of adaptive radiation, the class Insecta split into at least three distinctive branches. Each branch represents a group of orders that share structural or developmental similarities, and presumably, arose from a common ancestor. Insects in the first branch, the Orthoperoids, are relatively unspecialized. They all have chewing mouthparts and incomplete (hemimetabolous) development. A second branch, the Hemipteroids, also includes hemimetabolous insects but their mouthparts show varying degrees of specialization for scraping, rasping, or piercing/sucking. The third branch encompasses all insects that undergo complete metamorphosis (holometabolous development). These insects have four stages in their life cycle: egg, larva, pupa, and adult. Systematists group all holometabolous insects as Endopterygota (from the Greek "*endo*" meaning inner and "*pterygo*" meaning wing). This name reflects the fact that wing buds appear only during the pupal stage, arising internally from embryonic tissues.

The Orthopteroid Group

Orthopteroids have a very simple, unspecialized body-plan that retains many of the ancestral (pleisiomorphic) characteristics of ametabolous insects: abdominal cerci, chewing mouthparts, long multi-segmented antennae, and a distributed nervous system with numerous segmental ganglia.

Adults have four wings, although some species are secondarily wingless. The front wings (often called tegmina) are usually thickened or leathery. At rest, they cover and protect the hind wings. In flight, front and hind wings operate independently of one another (as in the Paleoptera). Hind wings are often enlarged near the base, providing a greater surface area for lift during flight. Most of the orthopteroids are rather weak or clumsy fliers.

There is extensive controversy over phylogenetic relationships within the orthopteroid complex. Although the fossil record contains many primitive neopterans, few systematists agree on how these extinct organisms are related to living orders and families. The ordinal status of modern-day orthopteroids is also the subject of much debate: "lumpers" are inclined to group all of these insects into five or six orders, whereas "splitters" divide them into as many as ten different orders.

Under the classification scheme we have chosen to use in this book each major ecological group is given ordinal status. This may please the "splitters" but it probably gives a false impression that the evolutionary history of these organisms is more diverse than it really is. In fact, there is strong justification for combining some of these orders, and we will try to emphasize these groupings in the following paragraphs.

Plecoptera

The first orthopteroid insects were scavengers and/or herbivores. From a physical standpoint, they were probably very similar to members of the present-day order Plecoptera. These insects, commonly known as stoneflies, are generally regarded as the earliest group of Neoptera. They probably represent an evolutionary "dead end" that diverged well over 300 million years ago. Immature stoneflies are aquatic nymphs (naiads). They usually live beneath stones in fast-moving, well-aerated water. Oxygen diffuses through the exoskeleton or into tracheal gills located on the thorax, behind the head, or around the anus.

Most species feed on algae and other submerged vegetation, but two families (Perlidae and Chloroperlidae) are predators of mayfly nymphs (Ephemeroptera) and other small aquatic insects. Adult stoneflies are generally found on the banks of streams and rivers from which they have emerged. They are not active fliers and usually remain near the ground where they feed on algae or lichens. In many species, the adults are short-lived and do not have functional mouthparts. Stoneflies are most abundant in cool, temperate climates.

Embioptera

The order Embioptera (webspinners or embiids) is another group within the orthopteroid complex that probably appeared early in the Carboniferous period. Many insect taxonomists believe webspinners may represent another evolutionary "dead end" that diverged about the same time as Plecoptera. Determining phylogenetic relationships for this group is unusually difficult because the Embioptera have a number of adaptations not found in any other insects. The tarsi of the front legs, for example, are enlarged and contain glands that produce silk. No other group of insects, fossil or modern, have silk-producing glands in the legs. The silk is used to construct elaborate nests and tunnels under leaves or bark. Webspinners live gregariously within these silken nests, feeding on grass, dead leaves, moss, lichens, or bark. Nymphs and adults are similar in appearance. Embiids rarely leave their silken tunnels; a colony grows by expanding its tunnel system to new food resources. Well-developed muscles in the hind legs allow these insects to run backward through their tunnels as easily as they run forward. Only adult males have wings. Front and hind wings are similar in shape and unusually flexible; they fold over the head when the insect runs backward through its tunnels.

Blood (hemolymph) is pumped into anterior veins to stiffen the wings during flight. In Embioptera, the mouthparts are directed forward (prognathous) rather than downward as in other primitive orthopteroids. This may simply be an adaptation for life in a tunnel, or as some taxonomists have suggested, it may mean that Embioptera are really more closely related to earwigs (order Dermaptera). Most Embioptera are tropical or subtropical.

Blattodea

The ancestral prototype for the main line of orthopteroid evolution was probably an insect very similar in appearance to a cockroach. Paleobiologists refer to this ancestral lineage as the Protoblattodean line. It probably dates from the early Carboniferous period, around 360 million years ago. In fact, fossil cockroaches found in late Carboniferous rock are remarkably similar to species living today.

In our scheme of classification, all modern cockroaches are grouped in one order, the Blattodea (or Blattaria). "Lumpers" often put them together with praying mantids (in the order Dictyoptera) or include them as a suborder of Orthoptera. The cockroaches, often known as "waterbugs" are scavengers or omnivores. They are most abundant in tropical or subtropical climates, but they also inhabit temperate and boreal regions. They are commonly found in close association with human dwellings where they are considered pests. Cockroaches have an oval, somewhat flattened body that is well-adapted for running and squeezing into narrow openings.

Rather than flying to escape danger, roaches usually scurry into cracks or crevices. Much of the head and thorax is covered and protected dorsally by a large plate of exoskeleton (the pronotum).

When cockroaches lay eggs, the female's reproductive system secretes a special capsule around her eggs. This structure, known as an öotheca, may be dropped on the ground, glued to a substrate, or retained within the female's body. Production of an öotheca is a special adaptation found only in the cockroaches and praying mantids. This similarity suggests a close phylogenetic relationship between these groups and explains why some taxonomists prefer to lump them into a single order (Dictyoptera).

Mantodea

From an ecological standpoint, cockroaches and mantids could not be more different: roaches are nocturnal scavengers, mantids are diurnal predators — in fact, they are the largest group of predators in the entire orthopteroid complex. Mantids.

Order Mantodea, have elongate bodies that are specialized for a predatory lifestyle: long front legs with spines for catching and holding prey, a head that can turn from side to side, and cryptic coloration for hiding in foliage or flowers. Mantids are most abundant and most diverse in the tropics; there are only 5 species commonly collected in the United States and 3 of these have been imported from abroad.

Isoptera

The termites, order Isoptera, are another group of insects that appear to be closely related to cockroaches. This conclusion is based on behavioral and ecological similarities between termites and wood roaches (members of the family Cryptocercidae). These cockroaches live in fallen timber on the forest floor, feeding on wood fibres which are then digested by symbiotic microorganisms within their digestive systems. They live in small family groups where each female provides care for her young offspring. Termites and wood roaches are thought to be close relatives because they both occupy similar habitats, share the same type of food resources, have the same intestinal symbionts, and provide care for their offspring.

Termites are the only hemimetabolous insects that exhibit true social behavior. They build large communal nests that house an entire colony. Each nest contains adult reproductives (one queen and one king) plus hundreds or thousands of immatures that serve as workers and soldiers. Like cockroaches and mantids, the termites are most abundant in tropical and subtropical climates.

In Blattodea, Mantodea, and Isoptera, wing movement (particularly the downstroke) is largely dependent on muscles attached to the base of the wing (direct flight muscles). But in another branch of the Protoblattodean lineage, direct flight muscles are smaller and more of the power for flight is provided by indirect flight muscles (located in the thorax but not attached directly to the wings). At least two extinct orders (Protorthoptera and Protelytroptera) appear to be part of this second branch which also includes all the rest of the modern-day orthopteroid orders: Orthoptera, Phasmatodea, Dermaptera, Grylloblattodea, and perhaps Zoraptera and Mantophasmatodea.

Orthoptera

Living members of this order are terrestrial herbivores with modified hind legs that are adapted for jumping. Slender, thickened front wings (tegmina) fold back over the abdomen to protect membranous, fan-shaped hind wings. Many species have the ability to make and detect sounds. Orthoptera is one of the largest and most important groups of plant-feeding insects.

Although their phylogeny is not clear, all other members of the orthopteroid complex are probably sister groups to the Orthoptera. These include the earwigs (order Dermaptera), leaf and stick insects (order Phasmatodea), rock crawlers (order Grylloblattodea), gladiators (order Mantophasmatodea), and zorapterans (order Zoraptera). Nearly all of these insects are herbivores or scavengers. In earwigs and stick insects, the chewing mouthparts are directed forward (prognathous) as in Embioptera; in rock crawlers, gladiators, and zorapterans, the mouthparts are directed downward (hypognathous) as in all other orthopteroids.

Phasmatodea

The leaf and stick insects (order Phasmatodea or Phasmida) are sometimes grouped as a family or suborder of Orthoptera. All species are herbivores. As the name "walkingstick" implies, most phasmids are slender, cylindrical, and cryptically colored to resemble the twigs and branches on which they live. Members of the family Timemidae (=Phyllidae) bear a strong resemblance to leaves: abdomens are broad and flat, legs have large lateral extensions, and coloration is primarily brown, green, or yellow. Most walkingsticks are slow-moving insects, a behavior pattern that is consistent with their cryptic lifestyle. In a few tropical species, the adults have well-developed wings, but most phasmids are brachypterous (reduced wings) or secondarily wingless.

Stick insects are most abundant in the tropics where some species may grow to 30 cm (12 inches) in length. Females do not have a well-developed ovipositor so they cannot insert their eggs into host plant tissue like most other Orthoptera. Instead, the eggs are dropped singly to the ground, sometimes from great heights.

Dermaptera

Earwigs (order Dermaptera) are mostly scavengers or herbivores that hide in dark recesses during the day and become active at night. They feed on a wide variety of plant or animal matter. A few species may be predatory. Females lay their eggs in the soil and may guard them until they hatch. In a few species, maternal care even extends through the first two instars. Nymphs are similar in appearance to adults, but lack wings. The front wings are short, thick, and serve as protective covers for the hind wings. Hind wings are large, fan-shaped and pleated. They fold (both length-wise and cross-wise) to fit beneath the front wings when not in use. Some species are secondarily wingless. In most earwigs, the **cerci** at the end of the abdomen are enlarged and thickened to form pincers (forceps). These pincers are used in grooming, defense, courtship, and even to help fold the hind wings.

The Dermaptera contains three suborders. Most species belong to the Forficulina. The other two groups, Arixeniina and Hemimerina, live in close association with mammals. The former (five species) live on Asian bats and the latter (eleven species) live on African rodents. All of these insects are adapted for a parasitic or semi-parasitic lifestyle: they are secondarily wingless and the cerci are not well-developed into pincers. Members of the Arixeniina give birth to live nymphs (vivipary).

Grylloblattodea

The rock crawlers (order Grylloblattodea) are a small and obscure group of insects found only at high elevations in the mountains of China, Siberia, Japan, and western United States and Canada. Cave-dwelling species have been found in Korea and Japan. These omnivorous insects scavenge for food on the surface of snowfields, under rocks, or near melting ice.

They are active only at cold temperatures and move downward toward permafrost during warm seasons. As their name implies, rock crawlers have a blend of physical characteristics from both crickets (gryllo-) and cockroaches (blatta-). Some taxonomists include these insects as a suborder or family within Orthoptera. Others believe these insects are the only survivors of a primitive lineage that gave rise to other orthopteroid orders.

Mantophasmatodea

Living members of this group have been found only in the Brandberg and Erongo Mountains of Namibia and the Western Cape Province of South Africa. These insects appear to be nocturnal predators. They live within rock crevices, hide in clumps of grass, and prey on spiders and other small insects. As their order name suggests, they seem to exhibit a blend of the physical and ecological characteristics found in praying mantids (Mantodea) and walkingsticks (Phasmatodea).

Zoraptera

Zoraptera, the final order within the orthopteroid complex, is probably the most controversial in terms of its phylogenetic position within the class Insecta. In many respects, the Zoraptera are typical orthopteroids: they have chewing mouthparts, unsegmented cerci, and a striking resemblance to termites. But other features are more typical of insects in the hemipteroid complex: the front wings (when present) are larger than the hind wings and have reduced venation, the nervous system has a reduced number of abdominal ganglia, and there are very few Malpighian tubules (excretory structures) in the digestive system.

This blend of orthopteroid and hemipteroid characteristics has led some entomologists to propose that Zoraptera represent a link between the two evolutionary lineages. Others reject this idea and claim that Zoraptera should be grouped with the protoblattodean lineage, near cockroaches, termites, and mantids. Still others argue that Zoraptera is a descendant of the protelytropteran lineage and therefore related to Dermaptera and Grylloblattodea. Regardless of phylogenetic placement, it seems likely that some of Zoraptera's derived (apomorphic) characteristics are the result of convergent evolution.

Members of the order Zoraptera are small (less than 4 mm) and usually found in rotting wood, under bark, or in piles of old sawdust. They live in small aggregations and appear to scavenge on spores and mycelium of fungi, or occasionally, on mites and other small arthropods. Little more is known about their biology. Some Zoraptera are blind, pale in colour, and wingless, while other members of the same species may be darkly pigmented with compound eyes and wings. The winged individuals are rather uncommon; they may be dispersal forms. The wings break off easily near the base, leaving only stubs.

THE HEMIPTEROID GROUP

Although it contains fewer orders, the hemipteroid branch of the insect's phylogenetic tree encompasses nearly three times as many species as the orthopteroid branch and has far greater ecological and economic impact. All of the insects classified here exhibit various "reductions" or "simplifications" from the primitive body-plan found in typical orthopteroids. Cerci, for example, are entirely absent in all living hemipteroids (some taxonomists classify this group as the superorder Acercaria, meaning without cerci). Other "reductions" occur in wing venation, in the number of tarsal segments (not more than three), in the number of Malpighian tubules (not more than six), and in the number of ganglia present in the ventral nerve cord (not more than five).

Perhaps the most significant hallmark of the hemipteroid lineage is a progressive development of haustellate mouthparts, variously adapted for scraping food from a substrate or gathering it by suction in liquid or semi-liquid form. Mouthparts of this type are distinctive because one or more of their structural components is elongated to form a rod or a stylet. In their most specialized form, as in the order Hemiptera, haustellate mouthparts form a tubular rostrum (also called a beak or a proboscis) with feeding stylets that can pierce the tissue of a plant or animal host and suck out the liquid contents. Enlargement of cibarial muscles (along the anterior surface of the pharynx) usually accompanies the development of haustellate mouthparts. Contraction of these muscles creates the bellows-like suction that draws food up through the proboscis.

Entomologists generally agree that members of the Hemipteroid complex arose from a common ancestor during the Carboniferous period, around 290 million years ago. Adaptive radiation within this lineage eventually led to the evolution of two sister groups one giving rise to the present-day booklice, barklice, and parasitic lice (Psocoptera and Phthiraptera); the other giving rise to thrips, true bugs, aphids, leafhoppers, etc. (Thysanoptera and Hemiptera).

Psocoptera

The order Psocoptera (also known as Corrodentia) contains the booklice and barklice. These insects are often regarded as the most primitive hemipteroids alive today because their mouthparts show the least modification from a primitive mandibulate condition. In fact, only the lacinia (a subdivision of the maxilla) has become a separate, rod-like structure that is pushed against the substrate as a brace while the mandibles scrape off surrounding food particles. The pharynx and hypopharynx are also modified for grinding food in a mortar-and-pestle arrangement.

Barklice generally live in moist terrestrial environments (in leaf litter, beneath stones, on vegetation, or under the bark of trees) where they forage on algae, lichens, fungi, and various plant products. They may grow to 10 mm in length and are frequently winged during the adult stage. Some species are gregarious. They often live in small colonies beneath a gossamer web spun with silk from their labial glands. Booklice are more common in human dwellings and warehouses. They are wingless and much smaller than barklice (less than 2 mm). Most species feed on stored grains, book bindings, wallpaper paste, fabric sizing, and other starchy products.

Although most Psocoptera are free-living, a few genera live in the nests of birds. They survive by feeding on residues of feathers or skin cells, but never on the birds themselves. Most entomologists suspect that true lice (ectoparasities of birds and mammals) evolved directly from these commensal barklice. Even today, it is possible to find a gradual progression of species with increasing dependence on vertebrate hosts. A close phylogenetic relationship between barklice and parasitic lice is also supported by similarities in the structure of mouthparts (particularly the hypopharynx).

Phthiraptera

There is a continuing debate among entomologists regarding the ordinal grouping of parasitic lice. "Splitters" divide these insects into biting lice (order Mallophaga) and sucking lice (order Anoplura). The distinction is based primarily on the presence or absence of mandibles that are suitable for biting and chewing. "Lumpers" include all parasitic lice in a single order (the Phthiraptera). As justification, they cite numerous similarities in structure and ecology. In this textbook we have chosen to "lump the lice". This decision reflects the growing consensus among many entomologists that sucking lice are specialized descendents of biting lice.

All Phthiraptera are wingless external parasites of birds and mammals. The biting lice probably evolved first on birds, feeding on feathers and dead skin cells. But sometime during the Cretaceous period (less than 135 million years ago) biting lice expanded their host range to include certain groups of mammals. A few of these lice (suborder Rhynchophthirina) developed the habit of breaking their host's skin and feeding on its blood. This lineage presumably gave rise to the sucking lice (suborder Anoplura), all of which are blood-feeding ectoparasites of placental mammals.

Unlike many other ectoparasites, the Phthiraptera cannot survive long if separated from the body of their host. Eggs (called nits) are glued directly to the hair or feathers and nymphs feed on the parental host. Since lice have no wings, dispersal to new host animals is limited to occasions when members of the host species come into direct contact with each other. This close interspecific association means that most lice are limited to a very narrow host range—often only a single species.

Thysanoptera

A second branch of the hemipteroid lineage includes insects in which the haustellate mouthparts contain feeding stylets derived from both the mandibles and maxillae. In Thysanoptera (thrips), there are three needle-like stylets that form an asymmetrical feeding apparatus housed within a conical mouth opening. The stylets (two maxillae and one mandible) pierce and lacerate host tissue. Macerated food is then drawn into the mouth by suction from the cibarial pump.

Thrips are generally small insects (under 3 mm). Most species feed on plant tissues (often in flower heads), but some are predators of mites and various small insects (including other thrips). Many species are parthenogenetic. Adults may be winged or wingless. When present, the wings are slender and rod-like with a dense fringe of long hairs.

Although Thysanoptera are hemimetabolous, many species undergo an extended metamorphosis in which the final immature stage is quiescent, non-feeding, and sometimes even enclosed in a silken cocoon. This developmental stage, usually called a "pupa", has aroused a great deal of speculation by some entomologists who claim that thrips represent an "intermediate" stage between hemi- and holometabolous development. A close examination of the thysanopteran "pupa", however, reveals that it does not undergo any internal transformation. Without additional evidence to support a phylogenetic link to the Holometabola, it would appear that this "pupal stage" may be nothing more than a curious coincidence of convergent evolution.

Hemiptera

Hemiptera is the final order in the hemipteroid complex and by far the largest, both in number of species and in ecological importance. All of the insects classified as Hemiptera have a highly specialized rostrum (proboscis) in which all structural elements are elongated to form a tubular feeding channel. The mandibles and maxillae are long and thread-like. All four of these feeding stylets interlock to form a flexible feeding tube that is no more than 0.1 mm in diameter yet contains both a food channel and a salivary channel. The stylets are enclosed within a protective sheath (the labium) that shortens or retracts during feeding.

Not long after these piercing/sucking mouthparts evolved, the hemipteran lineage seems to have split into two sister groups. In one group, Homoptera, the rostrum is relatively short (1-3 segments) and emerges from near the ventral posterior margin of the head. These insects (e.g., leafhoppers, cicadas, aphids, etc.) are said to have opistognathous mouthparts (from the Greek "*opisto-*" meaning backward and "*gnatha*" meaning mouth). Although some Homoptera are secondarily wingless, the majority have membranous or uniformly textured wings that fold tent-like over the body at rest.

In members of the second group, Heteroptera, the rostrum is relatively long (3-4 segments) and arises near the front or lower front of the head (prognathous or hypognathous).

These insects are known as the "true bugs". They have very distinctive front wings, called hemelytra, in which the basal half is leathery and the apical half is membranous. At rest, these wings cross over one another to lie flat along the insect's back.

For many years, the ordinal status of Heteroptera and Homoptera has been a source of great controversy among entomologists. Although "splitters" still prefer to divide them into two separate orders, a majority of taxonomists now classify Heteroptera and Homoptera as suborders within Hemiptera. Since the order is large and diverse, we will often use suborder names when referring to characteristics that are not common to both groups.

The Heteroptera is an unusually diverse assemblage of insects. Members of this suborder have become adapted to a broad range of habitats—terrestrial, aquatic, and semi-aquatic. Terrestrial species are often associated with plants. They feed in vascular tissues or on the nutrients stored within seeds. Other species live as scavengers in the soil or underground in caves or ant nests. Still others are predators on a variety of small arthropods.

Heteroptera

A few species even feed on the blood of vertebrates. Bed bugs, and other members of the family Cimicidae, live exclusively as ectoparasites on birds and mammals (including humans). Aquatic Heteroptera can be found on the surface of both fresh and salt water, near shorelines, or beneath the water surface in nearly all freshwater habitats. With only a few exceptions, these insects are predators of other aquatic organisms.

Members of the suborder Homoptera feed by withdrawing sap from vascular plants. But beyond this, it is difficult to generalize about the biology of these insects. Cicadas are the largest members of the suborder. As nymphs, they live underground and feed on the roots of trees and shrubs. Some species complete development in as little as four years, but others have a 13- or 17-year life cycle. In contrast, the aphids are tiny, soft-bodied insects with multiple generations per year. Many species have complex life cycles involving more than one host plant. Winged and wingless forms of the same species may develop at different times of the year. Asexual reproduction (parthenogenesis) is common and males are unknown in some species. The scale insects are even more specialized. During much of their life cycle, they remain immobile, living beneath an impervious cover of wax or cuticle that they secrete over themselves. Legs and antennae often disappear after the first molt. Only newly hatched nymphs and adult males bear any resemblence to other insects. Females grow to sexual maturity, mate, produce off-spring, and die without ever leaving their protective cover.

In most of the Homoptera, a portion of the digestive system is modified into a filter chamber. This structure allows the insects to ingest and process large volumes of plant sap. Excess water, sugars, and certain amino acids bypass most of the midgut and are shunted directly into the hindgut for excretion as honeydew. Only a small volume of filtered plant sap passes through the midgut for digestion and absorption. Many species of ants are attracted by the honydew and provide provide care and protection for the homopterans in exchange for the honeydew they excrete.

THE ENDOPTERYGOTE GROUP

Complete metamorphosis (holometabolous development), is the most distinctive characteristic of all endopterygote orders. These insects have four developmental stages in the life cycle: egg, larva, pupa, and adult (imago). The larval stage is a period of active feeding and growth. The pupal stage is a period of reconstruction: larval tissues are broken down (histolyzed) and rebuilt according to the adult body plan. The adult stage is a period of dispersal and reproduction.

The ancestors of today's holometabolous insects may have first appeared near the end of the Carboniferous period (around 290 million years ago) — probably not long after Hemipteroids and Orthopteroids diverged from their ancestral forms. From an evolutionary point of view, this event is nearly as significant as the development of wings because it marks the point at which adaptions of immatures (for feeding and growth) became functionally independent from adaptations of the adults (for reproduction and dispersal). Unlike hemimetabolous insects in which the immature structures (e.g., legs, eyes, antennae, etc.) must also serve the adults, holometabolous insects acquire a completely new body during the pupal stage. In effect, the insect is pre-programmed with two sets of genetic instructions, one for the larval stage and a different one for the adult stage. This complete transformation makes it possible for larvae and adults to respond to selective pressures in different ways, to develop independent adaptations, and even to evolve very different lifestyles.

The selective advantage of holometabolous development is under-scored by today's dominance of these insects in the world's fauna. Although endopterygote orders represent only about 1/3 of the living insect orders, they contain over 4/5 of the known insect species. In fact, the four largest orders of insects, based on number of species, are all members of the Holometabola. Mother Nature likes good breeding!

Despite a reasonably good fossil record, it is still impossible to trace any endopterygote lineage all the way back to its Permian ancestors. Consequently, proposed phylogenetic relationships among the Holometabola remain largely speculative. Our current theories are "educated guesses"; they are only as good as the assumptions upon which they are based.

Most taxonomists assume that the first holometabolous insects were relatively "unspecialized": mandibulate mouthparts, four similar wings with heavy venation, five-segmented tarsi, and active immature stages. Because Mecoptera (scorpionflies) and Neuroptera (lacewings and their relatives) are the two endopterygote orders that most nearly fit this description, either one might represent the ancestral body plan. In fact, both orders may have a legitimate claim to this distinction because fossils of each order are among the first to appear in rocks of the Permian period. It is certainly reasonable to suggest that two sister groups diverged from a primitive endopterygote ancestor and that all living members of the Holometabola descended from one of these two lineage). If this interpretation is not correct, then there may be as many as six separate lines of evolution leading away from a common endopterygote ancestor.

One noticable trend in the phylogeny of endopterygote orders is a structural or functional reduction in the number of wings. The most primitive orders have two pairs of wings, all more or less similar in size, with independent musculature and asynchronous wing beat. In the more specialized orders, one or more pair of wings becomes vestigal (Strepsiptera and Siphonaptera), modified in function (Coleoptera and Diptera), or coupled together with other wings by means of hooks, hairs, or bristles to act as a single flight surface (Hymenoptera, Trichoptera, and Lepidoptera).

Neuroptera

The order Neuroptera includes the lacewings and antlions (suborder Planipennia), dobsonflies and alderflies (suborder Megaloptera) and snakeflies (suborder Raphidiodea). "Splitters" prefer to assign each of these groups to a separate order (Neuroptera, Megaloptera, and Raphidioptera, respectively), based on differences in structure and development. In this case, however, we have sided with the "lumpers" who argue that grouping these insects within a single order reflects the broader similarities in wing structure and venation shared by all adults, and the presence of grasping/biting mouthparts and a predatory lifestyle among all larvae.

The Megaloptera are always aquatic as immatures. They live under stones or submerged vegetation and feed on a variety of small aquatic organisms. Large species, often called hellgrammites, may require several years of growth to reach maturity. Adults usually remain near water, although they are attracted to lights at night. In most species, the adults live only a few days and rarely feed.

Except for larval spongillaflies (family Sisyridae) which feed on fresh-water sponges, all members of the of the suborders Planipennia and Raphidiodea are terrestrial. Antlion larvae live in the soil and construct pitfall traps to snare prey. Lacewing larvae are usually found in vegetation where they typically feed on aphids, mites, and scale insects. Snakefly larvae live in leaf litter or under bark and catch aphids or other soft-bodied prey. In most cases, the adults of these insects are also predators—the non-predatory species usually feed on nectar, pollen, or honeydew.

The larvae of antlions and lacewings have specialized mouthparts with large, sickle-shaped mandibles and maxillae that interlock to form pincers. Once impaled on these pincers, a prey's body contents are sucked out through hollow food channels running between the adjacent surfaces of the mandibles and maxillae.

As adults, all neuropterans have two pairs of membranous wings with an extensive pattern of veins and crossveins. At rest, the wings are folded flat over the abdomen or held tent-like over the body. Most species are rather weak fliers.

When they pupate, larvae of lacewings and antlions dig a small cavity in the soil and spin a loose silken cocoon around themselves. Many holometabolous insects exhibit similar behavior, but neuropterans are unusual because their silk is produced by Malpighian tubules (excretory organs) and spun from the anus. In contrast, most other endopterygote insects produce silk in modified salivary glands and spin it with mouthparts. Only one other order, the Coleoptera, makes silk in the same manner as Neuroptera.

Coleoptera

A phylogenetic link between Neuroptera and Coleoptera is further strengthened by the existence of Permian fossils that appear to share the characteristics of both orders. Some of these beetle-like insects evidently evolved as carnivores (suborder Aedephaga) while others were herbivores (suborder Polyphaga). Significant growth and diversification of the herbivore lineage occurred near the beginning of the Cretaceous period (100-130 million years ago) when flowering plants (angiosperms) first began to take over as the dominant form of plant life.

Coleoptera (beetles and weevils) is the largest order in the class Insecta. As adults, most beetles have a hard, dense exoskeleton that covers and protects most of their body surface. The front wings, known as elytra, are just as hard as the rest of the exoskeleton. They fold down over the abdomen and serve as protective covers for the large, membranous hind wings. At rest, both elytra meet along the middle of the back, forming a straight line that is probably the most distinctive characteristic of the order. During flight, the elytra are held out to the sides of the body where they provide a certain amount of aerodynamic stability.

Both larvae and adults have strong mandibulate mouthparts. As a group, they feed on a wide variety of diets, inhabit all terrestrial and fresh-water environments, and exhibit a number of different life-styles. Many species are herbivores — variously adapted to feed on the roots, stems, leaves, or reproductive structures of their host plants. Some species live on fungi, others burrow into plant tissues, still others excavate tunnels in wood or under bark.

Many beetles are predators. They live in the soil or on vegetation and attack a wide variety of invertebrate hosts. Some beetles are scavengers, feeding primarily on carrion, fecal material, decaying wood, or other dead organic matter. There are even a few parasitic beetles—some are internal parasites of other insects, some invade the nests of ants or termites, and some are external parasites of mammals.

Strepsiptera

Most beetle larvae that parasitize insects are free-living as adults. But the stylopids (order Strepsiptera) are thought to be a closely related group that have retained the parasitic lifestyle even as adults. Most Strepsiptera (also known as twisted-wing parasites) live as internal parasites of bees, wasps, grasshoppers, leafhoppers, and other members of the order Hemiptera. Only a few species that parasitize bristletails (Archeognatha) are known to be free-living in the adult stage.

Strepsiptera share so many characteristics with beetles that some entomologists classify them as a superfamily of Coleoptera. In fact, Strepsiptera and certain parasitic beetles (in the families Meloidae and Rhipiphoridae) are among the very few insects that undergo hypermetamorphosis, an unusual type of holometabolous development in which the larvae change body form as they mature. Upon emerging from their mother's body, the young larvae, called triunguloids, have six legs and crawl around in search of a suitable host. In species that parasitize bees or wasps, a triunguloid usually climbs to the top of a flower and waits for a pollinator. When a host arrives, the larva burrows into its body and quickly molts into a second stage that has no distinct head, legs, antennae or other insect-like features. These larvae grow and continue to molt inside the host's body cavity, assimilating nutrients from the blood and non-vital tissues. After pupating in the host, winged males emerge and fly in search of mates. An adult female remains inside her host, managing to attract and mate with a male while only a small portion of her body protrudes from the host's abdomen. Embryos develop within the female's body, and a new generation of triunguloid larvae begin their life cycle by escaping through a brood passage on the underside of her body.

Adult male Strepsiptera are strange-looking insects. The head is small, with protruding compound eyes that look like tiny raspberries. The antennae are multi-segmented and have up to three branches. Front wings are reduced to small, club-like structures; hind wings are very large and fan-shaped. In fact, Coleoptera and male Strepsiptera are the only holometabolous insects in which the hind wings are larger than the front wings and provide most of the lift during flight.

Mecoptera

At the same time beetles were diverging from their neuropteran ancestors, the Mecoptera were also diverging and specializing. In this lineage, front wings became the predominate flight surfaces. The hind wings were either modified in function (Diptera) or else reduced in size and linked together with the front wings (by means of hooks, hairs, or bristles) to form a single flight surface.

The Mecoptera (scorpionflies) are a curious group of terrestrial insects that usually live in moist sylvan habitats. Both larvae and adults are omnivorous. Mostly, they feed upon decaying vegetation and dead (or dying) insects. Larvae generally remain in the soil; they have chewing mouthparts and resemble caterpillars (Lepidoptera) or white grubs (Coleoptera). Most adults have an elongated head with slender, chewing mouthparts near the tip of a stout beak. Front and hind wings are similar in shape (occasionally reduced in size or absent), and often mottled with patches of color. The common name of this order (scorpionfly) refers to the distinctive appearance of male genitalia in members of the family Panorpidae: the terminal segments are enlarged and held recurved over the abdomen like the tail of a scorpion. Despite its appearance, the scorpionfly's tail is quite harmless.

Hanging scorpionflies, family Bittacidae, are predators of small flying insects. Their legs, especially the tarsi, are unusually long and slender. At the tip of each leg there is a single opposable claw. The adults hang from vegetation with their front legs and catch small flying insects with their middle and hind legs. These scorpionflies, which bear a striking resemblance to crane flies (Diptera: Tipulidae), may have developed from the same ancestral lineage that also give rise to the caddisflies (order Trichoptera) and the true flies (order Diptera).

Diptera

The order Diptera includes all the true flies. These insects are distinctive because their hind wings are reduced to small, club-shaped structures called halteres — only the membranous front wings serve as aerodynamic surfaces. The halteres, which vibrate during flight, work much like a gyroscope to help the insect maintain balance while it is in flight.

All Dipteran larvae are legless. They live in aquatic (fresh water), semi-aquatic, or moist terrestrial environments. They are commonly found in the soil, in plant or animal tissues, and in carrion or dung—almost always where there is little danger of desiccation. Some species are herbivores, but most feed on dead organic matter or parasitize other animals, especially vertebrates, molluscs, and other arthropods. In the more primitive families (suborder Nematocera), fly larvae have well-developed head capsules with mandibulate mouthparts. These structures are reduced or absent in the more advance suborders (Brachycera and Cyclorrhapha) where the larvae, known as maggots, have worm-like bodies and only a pair of mouth hooks for feeding.

Adult flies live in a wide range of habitats and display enormous variation in appearance and life style. Although most species have haustellate mouthparts and collect food in liquid form, their mouthparts are so diverse that some entomologists suspect the feeding adaptations may have arisen from more than a single evolutionary origin. In many families, the proboscis (rostrum) is adapted for sponging and/or lapping.

These flies survive on honeydew, nectar, or the exudates of various plants and animals (dead or alive). In other families, the proboscis is adapted for cutting or piercing the tissues of a host. Some of these flies are predators of other arthropods (e.g., robber flies), but most of them are external parasites (e.g., mosquitoes and deer flies) that feed on the blood of their vertebrate hosts, including humans and most wild and domestic animals.

Siphonaptera

Judging from their haustellate mouthparts, Siphonaptera (fleas) may have diverged from one of the dipteran lineages. As adults, all fleas are blood-sucking external parasites. Most species feed on mammals, although a few (less than 10%) live on birds. Only adult fleas inhabit the host's body and feed on its blood. They are active insects with a hard exoskeleton, strong hind legs adapted for jumping, and a laterally flattened body that can move easily within the host's fur or feathers. Unlike lice, most fleas spend a considerable amount of time away from their host. Adults may live for a year or more and can survive for weeks or months without a blood meal.

Flea larvae are worm-like (vermiform) in shape with a sparse covering of bristles. They rarely live on the body of their host. Instead, they are usually found in its nest or bedding where they feed as scavengers on organic debris (including adult feces). In general, flea larvae can survive more arid conditions than most fly larvae. After a larval period that includes two molts, fleas pupate within a thin silken cocoon. Under favorable conditions, the life cycle can be completed in less than a month.

Trichoptera

The order Trichoptera (caddisflies) is another likely descendant of the Mecopteran lineage. Adults are mostly nocturnal, weak-flying insects that are often attracted to lights. During the day, they hide in cool, moist environments such as the vegetation along river banks. The body and wings are clothed with long silky hairs (setae)—a distinctive characteristic of the order. In flight, the hind wings are coupled to the front wings by specially curved hairs. At rest the wings are held tent-like over the abdomen. Many caddisflies have reduced or vestigal mouthparts. Few species have actually been observed feeding, and most adults are relatively short-lived.

All caddisfly larvae live in aquatic environments; they may be herbivores, scavengers, or predators. In most cases, the predatory species are free-living or spin silken structures in the water (webs or tunnels) to entrap prey. The scavengers and herbivores live within protective "cases" which they build from their own silk and stones, twigs, leaf fragments, or other natural materials.

Case design and construction is distinctive for each family or genus of caddisfly. The case is usually portable, dragged around like a snail shell as the insect moves, and held in place by a pair of hooked prolegs at the tip of the abdomen. Most species have thread-like abdominal gills and get oxygen from water that circulates inside the case. All larval growth and development (including pupation) occurs within the case.

Lepidoptera

Lepidoptera (moths and butterflies) is the second largest order in the class Insecta. Adults are distinctive for their large wings (relative to body size) which are covered with minute overlapping scales. Most entomologists believe that these scales are structurally related to the hair (setae) covering adult caddisflies. Lepidopteran wing scales often produce distinctive color patterns that play an important role in courtship and intraspecific recognition. In flight, front and hind wings are linked together by a bristle (frenulum) or a membranous flap (jugum) so both wings move up and down in synchrony.

Although moths probably diverged from caddisflies in the early Triassic period, about 230 million years ago, adults in a few primitive families (e.g.. Micropterygidae) still retain evidence of chewing mouthparts. In all other lepidopteran families, the mouthparts are vestigal or form a tubular proboscis that lies coiled like a watch spring beneath the head. This proboscis is derived from portions of the maxillae. It uncoils by hydrostatic pressure and acts as a siphon tube for sipping liquid nutrients, such as nectar, from flowers and other substrates.

From a taxonomic standpoint, the distinction between moths and butterflies is largely artificial—some moths are more similar to butterflies than to other moths. As a rule, butterflies are diurnal, brightly colored, and have knobs or hooks at the tip of the antennae. At rest, the wings are held vertically over the body. In contrast, most (but not all) moths are nocturnal. They are typically drab in appearance, and have thread-like, spindle-like, or comb-like antennae. At rest, their wings are held horizontally against the substrate, folded flat over the back, or curled around the body.

Nearly all lepidopteran larvae are called caterpillars. They have a well-developed head with chewing mouthparts. In addition to three pairs of legs on the thorax, they have two to eight pairs of fleshy abdominal prolegs that are structurally different from the thoracic legs. Most lepidopteran larvae are herbivores; some species eat foliage, some burrow into stems or roots, and some are leaf-miners.

Hymenoptera

The Hymenoptera (sawflies, ants, wasps, and bees) is a distinctive group of insects that has relatively few advanced characteristics in common with other endopterygote groups. Although the fossil record for Hymenoptera dates back only to the Triassic period (220-230 million years ago), we suspect that these insects first appeared in the Permian period (225-290 million years ago) as an early offshoot of the mecopteran lineage. This interpretation is supported by the fact that sawflies and horntails, the most primitive members of the order, produce silk from modified salivary glands, have abdominal prolegs in the larval stage, and show reduction in size of the hind wings. More advanced groups (wasps, bees, and ants) appeared much later in the fossil record (mid-Jurassic) and probably co-evolved with flowering plants during the Cretaceous period.

As a rule, members of the order Hymenoptera can be regarded as ecological specialists. Most species are rather narrowly adapted to specific habitats and/or specific hosts. Their remarkable success as a taxon probably has more to do with their immense range of behavioral adaptation rather than any physical or biochemical charcteristic. The Hymenoptera is the only order besides the Isoptera (termites) to have evolved complex social systems with division of labour.

Herbivory is common among the primitive Hymenoptera (suborder Symphyta), in the gall wasps (Cynipidae), and in some of the ants and bees. Most other Hymenoptera are predatory or parasitic. The large hunting wasps are agile predators that catch and paralyze insects (or spiders) as food for their offspring. The greatest diversity, though, is found among the many families of parasitoid wasps whose larvae feed internally on the living tissues of other arthropods (or their eggs). These insects eventually kill their host, but not before completing their own larval development within its body. Despite their small size and characteristically narrow host range, these wasps are highly abundant and exert a tremendous impact on the population dynamics of many other insect species.

Most of the Hymenoptera have relatively unspecialized mandibulate mouthparts. An exception is found in the bees (superfamily Apidoidea) where the maxillae and labium are modified into a proboscis that works like a tongue to collect nectar from flowers. In these insects, the mandibles are used to gather or manipulate pollen and wax.

8

Lycopodium

INTRODUCTION

Lycopodium is a genus of clubmosses, also known as ground pines, in the family Lycopodiaceae, a family of fern-allies. They are flowerless, vascular, terrestrial or epiphytic plants, with widely-branched, erect, prostrate or creeping stems, with small, simple, needle-like or scale-like leaves that cover the stem and branches thickly. The fertile leaves are arranged in cone-like strobilli. Specialized leaves (sporophylls) bear reniform spore-cases (sporangia) in the axils, which contain spores of one kind only. These club-shaped capsules give the genus its name.

Lycopods reproduce sexually by spores. The plant has an underground sexual phase that produces gametes, and this alternates in the life cycle with the spore-producing plant. The prothallium developed from the spore is a subterranean mass of tissue of considerable size and bears both the male and female organs (antheridium and archegonia). However, it is more common that they are distributed vegetatively through above or below ground rhizomes.

There are approximately 200 species, with 37 species widely distributed in temperate and tropical climates, though they are confined to mountains in the tropics.

Species

- *Lycopodium aberdaricum* (central and southern Africa)
- *Lycopodium alboffii* (southernmost South America and the Falkland Islands)
- *Lycopodium alticola* (southwest China)
- *Lycopodium annotinum* (Interrupted Clubmoss; circumpolar north temperate)
- *Lycopodium assurgens* (Brazil Minas Gerais, Santa Catarina)
- *Lycopodium casuarinoides* (southeast Asia (Japan to Bhutan and Borneo)
- *Lycopodium centrochinense* (east Asia' central China to India and the Philippines)
- *Lypocodium cernuum* creeping club moss (lowland mixed forest) — found along bush margins or disturbed ground; height approximately 400 mm
- *Lycopodium clavatum* (Stag's-horn Clubmoss; subcosmopolitan
- *Lycopodium confertum* (southern South America and the Falkland Islands)
- *Lycopodium dendroideum* (northern North America)
- *Lycopodium deuterodensum*, tree club moss (eastern Australia, New Caledonia, New Zealand) — has appressed leaves; height approximately 600 mm
- *Lycopodium diaphanum* (Tristan da Cunha)
- *Lycopodium dubium* (cold temperate and subarctic Europe and Asia; treated as a synonym of *L. annotinum* by some authors)
- *Lycopodium fastigiatum* (southeastern Australia, New Zealand)
- *Lycopodium gayanum* (south-central Chile and adjacent westernmost Argentina)

- *Lycopodium hickeyi* (northeastern North America)
- *Lycopodium hygrophilum* (New Guinea)
- *Lycopodium interjectum* (southwest China (Sichuan)
- *Lycopodium japonicum* [eastern Asia (Japan west and south to India and Sri Lanka)]
- *Lycopodium juniperoideum* [northeast Asia (central Siberia southeast to Taiwan)]
- *Lycopodium jussiaei* (northern South America, Caribbean)
- *Lycopodium lagopus* (circumpolar arctic and subarctic)
- *Lypocodium lucidulum*, shining club moss (North America) — occurs in wet woods and among rocks; has no distinct strobili; bears its spore capsules at the bases of leaves scattered along the branches
- *Lycopodium magellanicum* (South and Central America (Andes), southern Atlantic Ocean and southern Indian Ocean islands)
- *Lycopodium minchegense* [southeast China (Fujian)]
- *Lycopodium obscurum* (northeast North America, northeast Asia)
- *Lycopodium paniculatum* (southern South America (Andes)
- *Lycopodium papuanum* (New Guinea)
- *Lycopodium pullei* (New Guinea)
- *Lycopodium scariosum* (southeastern Australia, New Zealand, Borneo (Mount Kinabalu)
- *Lycopodium selago* (uplands of western Europe)
- *Lycopodium simulans* [southwest China (Yunnan)]
- *Lycopodium spectabile* (Java)
- *Lycopodium subarcticum* (northeast Siberia)
- *Lycopodium taliense* [southwest China (Yunnan)]

- *Lycopodium venustulum* (Hawaii, Western Samoa, Society Islands)
- *Lycopodium vestitum* (northwest South America (Andes)
- *Lycopodium volubile*, climbing club moss (southwest Pacific Ocean islands (New Zealand north to Java), Australia (Queensland) — found along bush margins and disturbed ground; has a creeping habit and can climb up vegetation
- *Lycopodium zonatum* (southeast Tibet)

The term *Lycopodium* is also used to describe the yellowish, powdery spores of certain club mosses, especially Lycopodium clavatum, used in the past in fireworks, fingerprint powders, as a covering for pills and explosives. The term "Lycopodium mask" is sometimes used to describe a type of flamethrower-mask worn by some music bands or artists on stage, such as Rammstein, most notably on the song Feuer frei!, featured in the movie xXx. Belisha (band) have got themselves banned when using the substance in flamethrowers. In Physics experiments, the powder is also used to make sound-waves in air visible for observation and measurement.

Lycopodium powder is also used to make a pattern of electrostatic charging visible. For example, Chester Carlson used lycopodium powder in his early experiments to demonstrate xerography.

It is also used as an ice cream stabilizer.

LYCOPODIOPSIDA

Lycopodiopsida is a class of plants often loosely grouped as the fern allies, and includes the clubmosses. Lycopodiopsida traditionally included all the clubmosses, including Selaginella and Isoetes.

Clubmosses are thought to be structurally similar to the earliest vascular plants, with small, scale-like leaves, homosporous spores borne in sporangia at the bases of the leaves, branching stems (usually dichotomous), and generally simple form.

The class Lycopodiopsida as interpreted here contains a single living order, the Lycopodiales, and a single extinct order, the Drepanophycales.

ORDER LYCOPODIALES

The classification of this group has been unsettled in recent years and a consensus is yet to emerge. Older classifications took a very broad definition of the genus *Lycopodium* that included virtually all the species of *Lycopodiales*. The trend in recent years has been to define *Lycopodium* more narrowly and to classify the other species into several genera, an arrangement that has been supported by both morphological and molecular data and adopted in numerous revisions and flora treatments. These genera fall into two distinct groups but there is as yet no consensus as to whether to recognize them in a single family, Lycopodiaceae, or to separate them into two families: a more narrowly defined Lycopodiaceae and Huperziaceae.

Lycopodiaceae, as narrowly defined, comprises the extant genus, *Lycopodium*, which includes the Wolf's-foot clubmoss, Lycopodium clavatum, Ground-pine, Lycopodium obscurum, Southern ground-cedar, Lycopodium digitatum, and other species. Also included are species of Lycopodiella, such as the Bog clubmoss, Lycopodiella inundata. Most of the Lycopodium favor acidic, sandy, upland sites, whereas most of the Lycopodiella favor acidic, boggy sites.

The other major group, the Family Huperziaceae, are known as the firmosses. This group includes the genus Huperzia, such as the Shining firmoss, Huperzia lucidula, the Rock firmoss, Huperzia porophila, and the Northern firmoss, Huperzia selago. This group also includes the odd, tuberous Australasian plant Phylloglossum, which was, until recently, thought to be only remotely related to the clubmosses. However, recent genetic testing has shown it to be very closely related to the genus Huperzia.

A powder known simply as lycopodium, consisting of dried spores of the common clubmoss, was used in Victorian theater to produce flame-effects. A blown cloud of spores burned rapidly and brightly, but with little heat. It was considered safe by the standards of the time.

FERN ALLY

Fern ally is a general term covering a somewhat diverse group of vascular plants that are not flowering plants and not true ferns. Like ferns, these plants disperse by shedding spores to initiate an alternation of generations. There were originally three or four groups of plants considered to be fern allies. In various classification schemes, these may be grouped as classes or divisions within the plant kingdom. The more traditional classification scheme is as follows (here, the first three classes are the "fern allies"):

- Kingdom: Plantae
- Division Tracheophyta (vascular plants)
- Class Lycopsida, (fern-allies) the clubmosses and related plants
- Class Sphenopsida or Equisetopsida, (fern-allies) the horsetails and scouring-rushes
- Class Psilopsida, (fern-allies) the whisk ferns
- Class Filices, the true ferns
- Class Spermatopsida (or sometimes as several different classes of seed-bearing plants)

A more modern or newer classification scheme is:

Psilotum nudum from the Psilopsida (whisk ferns)

- Kingdom Plantae
- Subkingdom Tracheobionta
- Division Lycopodiophyta
- Class Lycopodiopsida, the clubmosses
- Class Selaginellopsida, the spikemosses
- Class Isoetopsida, the quillworts and scale trees

- Division Pteridophyta
- Class Equisetopsida, the horsetails and scouring-rushes
- Class Psilotopsida, the whisk ferns, adders'-tongues and moonworts
- Class Pteridopsida, the true ferns
- Division Spermatophyta (or as several different divisions of seed-bearing plants)

Several groups of plants were considered "fern allies": the clubmosses, spikemosses, and quillworts in the Lycopodiophyta, the whisk ferns in Psilotaceae, and the horsetails in the Equisetaceae. More recent genetic studies have shown that the Lycopodiophyta are only distantly related to any other vascular plants, having radiated evolutionarily at the base of the vascular plant clade, while both the whisk ferns and horsetails are as much true ferns as are the Ophioglossoids and Marattiaceae. In fact, the whisk ferns and Ophioglossoids are demonstrably a clade, and the horsetails and Marattiaceae are arguably another clade. (Traditionally, three discrete groups of plants had been considered ferns: the adders-tongues, moonworts, and grape-ferns (Ophioglossales), the Marattiaceae, and the leptosporangiate ferns. The Marattiaceae are a primitive group of tropical ferns with a large, fleshy rhizome, and are now thought to be a sibling taxon to the main group of ferns, the leptosporangiate ferns.)

VASCULAR PLANT

Vascular plants (also known as tracheophytes or higher plants) are those plants that have lignified tissues for conducting water, minerals, and photosynthetic products through the plant. Vascular plants include the ferns, clubmosses, flowering plants, conifers and other gymnosperms. Scientific names for the group include Tracheophyta and Tracheobionta, ut neither name is very widely used.

Characteristics

Vascular plants are distinguished by two primary characteristics:

— Vascular plants have vascular tissues, which circulate resources through the plant. This feature allows vascular plants to evolve to a larger size than non-vascular plants, which lack these specialized conducting tissues and are therefore restricted to relatively small sizes.

— In vascular plants, the principal generation phase is the sporophyte, which is usually diploid with two sets of chromosomes per cell. By contrast, the principal generation phase in non-vascular plants is usually the gametophyte, which is haploid with one set of chromosomes per cell.

Water transport happens in either xylem or phloem: xylem carries water and inorganic solutes upward toward the leaves from the roots, while phloem carries organic solutes throughout the plant. Group of plants having lignified conducting tissue (xylem vessels or tracheids).

Nutrient Distribution

Nutrients and water from the soil and the organic compound produces in leaves are distributed to specific areas in the plant through the xylem and phloem. The xylem draws water and nutrients up from the roots to the upper sections of the plant's body, and the phloem conducts other materials, such as the glucose produced during photosynthesis, which gives the plant energy to keep growing and seeding.

The xylem consists of tracheids, which are dead hard-walled cells arranged to form tiny tubes to function in water transport. A tracheid cell wall usually contains the polymer lignin. The phloem however consists of living cells called sieve-tube members. Between the sieve-tube members are sieve plates, which have pores to allow molecules to pass through. Sieve-tube members lack such organs as nuclei or ribosomes, but cells next to them, the companion cells, function to keep the sieve-tube members alive.

Movement of nutrients, water, sugars and waste is effected by transpiration, conduction and absorption.

Transpiration

The most abundant compound in most plants is water, serving a large role in the various processes taking place. Transpiration is the main process a plant can call upon to move compounds within its tissues. The basic minerals and nutrients a plant is composed of remain, generally, within the plant. Water, however, is constantly being lost from the plant through its metabolic and photosynthetic processes to the atmosphere.

Water is transpired from the plants leaves via stomata, carried there via leaf veins and vascular bundles within the plants cambium layer. The movement of water out of the leaf stomata creates, when the leaves are considered collectively, a transpiration pull. The pull is created through water surface tension within the plant cells. The draw of water upwards is assisted by the movement of water into the roots via osmosis. This process also assists the plant in absorbing nutrients from the soil as soluble salts, a process known as absorption.

Absorption

Xylem cells move water and nutrient solutions upwards towards other plant organs from the roots and fine root hairs. Living roots cells actively absorb water in the absence of transpiration pull via osmosis creating root pressure. There are times when plants do not have transpiration pull, usually due to lack of light or other environmental elements. Water in the plant tissues may move to the roots to assist in passive absorption.

Conduction

Xylem and phloem tissues are involved in the conduction processes within plants. The movement of foods throughout the plant takes place mainly in the phloem. Plant conduction (food movement) is from an area of high food content, place of manufacture (photosynthesis) or storage, to a place of food utilisation, or from a point of manufacture to storage tissues. Mineral salts are translocated in the xylem tissues.

Index

A

Abnormal Phyllotaxy, 122
Aglaophyton, 146-147
Anodonta cygnea, 67
Arabidopsis thaliana, 116
Australian National Botanic Gardens, 40

B

Bacillus anthracis, 136
Bacillus subtilis, 134
Baragwanathia, 108
Bathurst Island, 68
Bryophyta, 81-87
  fascinating bryophyte, 86-87
  introduction, 81-83
  peristome, 83-86
  sporangium 83
Bryophyte, 8-80
  Campylopus clavatus, 52-68
  classification, 15-19
  Entosthodon apophysata, 36-38
  importance of mosses, 68-72
  introduction, 8-11
  leaves and peripheral structures, 72-73
  life cycle of mosses, 27-29
  monoicous organisms, 30-32
  moss reproduction, 32-35
  mosses, 13-15
  non-vascular plant, 11-13
  Polytrichum commune, 24-26
  reproduction and dispersal, 35-36
  saprophyte, 73-80
  sexual reproduction, 46-49
  – vs. vegetative, 38-41
  sphagnum, 21-24
  Takakia, 19-21
  Ulota phyllantha leaf gemma, 42-45
  vegetative reproduction, 49-52
Bryum argenteum, 39
Bryum, 86

C

C. elegans, 114
Campylopus clavatus, 40, 52-68
Carboniferous, 190
Ceratophyllum, 94
CLAVATA, 119
Clostridium tetani, 134
Common Haircap Moss, 24

D

Devonian, 109
DNA, 134, 135
Drosophila, 114

E

Earlier land plants, 143-155
- aglaophyton, 146-147
- introduction, 143-144
- lush life, 149-155
- rhynia, 147-149
- Silurian-devonian terrestrial algae and prolematica, 144-146

Eccremidium, 62
Entosthodon apophysata, 36-38
Equisetum, 109
Evolution and diversity, 156-200
- apterygote insects, 166-168
- *archaeognatha*, 168, 169
- arthropod evolution, 161
- biochemistry, 159-160
- blattodea, 178-179
- dermaptera, 181-182
- diptera, 195-196
- endopterygote group, 189-191
- ephemeroptera, 173-174
- first arthropods, 160-161
- gylloblattodea, 182
- hemiptera, 187-188
- hemipteroid group, 183-184
- heteroptera, 188-189
- hexapods, 166
- introduction, 156-157
- isopteran, 179-180
- lepidoptera, 198-199
- mantodea, 179
- mantophasmatodea, 182
- mecoptera, 195
- monophyletic hypothesis, 162-163
- myriapods, 164-165
- neopterous orders, 174-176
- neuropteran, 191-192
- orthoptera, 180-181
- orthopteriod group, 176-177
- paleopterous orders, 172-173
- phthiraptera, 185-186
- phylogenetic trees, 159
- polyphyletic hypothesis, 163-164
- psocoptera, 184-185
- siphonaptera, 196-197
- strepsiptera, 193-194
- subclass pterygota, 170-172
- thysanoptera, 186-187
- thysanura, 169
- thysanura, 169-170
- tools and techniques, 157-159
- trichoptera, 197-198

F

Fontinalis antipyretica, 16

G

Gylloblattodea, 182

H

Hepatica, 128
Hornworts, 88-123
- alternation of generations, 95-97
- ceratophyllum, 93-95
- diversity in meristem architectures, 119-120
- evolution of the meristem architecture, 120-121
- evolutionary history of plants, 98-118
- factors influencing leaf architectures, 121-122
- introduction, 88-93
- protists, 97-98

I

Introduction 1-7

K

KNOX, 119, 122

L

Lawrence Berkeley National Laboratory, 155
Lejeunea cardotii, 52
Lycopodium, 201-209
absorption, 209
conduction, 209
fern ally, 206-207
introduction, 201
lycopodiospida, 204-205
nutrient distribution, 208
order lycopodiales, 205-206
species, 202-204
transpiration, 209
vascular plant, 207

M

Marchantiophyta, 124-142
economic importance, 129
endospore, 133
formation and destruction, 135
introduction, 124-129
mosses and liverworts, 136-137
reactivation, 135-136
secondary endosymbiosis, 141-142
spore, 129-132
Megaceros, 89
Mother Nature, 190

N

Nostoc, 88
Notothylacites filiformis, 93

O

Olytrichum commune, 24

P

Phloem, 208
Photosynthesis, 209
Pilobolus, 132
Pilotrichella, 54
Pleurophascum, 58
Pohlia nutans, 52
Polytrichum commune, 24-26
Polytrichum, 16, 86
Psilotum, 148
Pteropus conspicillatus, 66

R

Rhynia, 147-149
Rhynie Chert, 148

S

Schistostega pennata, 64, 67
Selaginella, 108
Sphagnum, 7
Sphangnopsida, 16

T

Tortula muralis, 39

U

Ulota phyllantha leaf gemma, 42-45
University of Washington, 155
Utah State University, 155

X

Xenopus, 114
Xylem, 208

Z

Zebrafish, 114

❑❑❑